DIVERSITY OF DESERT LIFE

J. L.CLOUDSLEY-THOMPSON
J.A. Ph. D. (Cantab) D.Sc (Lond.)
Hon D.Sc (Khartoum), F I Biol, FWAAS,
Emeritus Professor of Zoology, University of London

SCIENTIFIC PUBLISHERS

Published by:

SCIENTIFIC PUBLISHERS (INDIA)
5A, New Pali Road, P.O. Box 91
Jodhpur 342 001, India
E-mail: info@scientificpub.com
Website: www.scientificpubonline.com

Print: 2021

ISBN: 978-81-72330-50-7
eISBN: 978-93-89061-70-3

Printed in India

CONTENTS

1

INTRODUCTION

THE DESERT ENVIRONMENT

'Whatever causes diversity ..... on the surface of our planet - mountains, great lakes, grassy steppes and even deserts', wrote Alexander von Humboldt in the 1800's, 'impresses some mark or character on .... its inhabitants'. von Humboldt was really thinking of the earth's human inhabitants, but his observation applies perhaps even more strongly to the totality of the flora and fauna. The exigencies of life in the desert environment have, indeed, resulted in the selection of a diversity of adaptations, both morphological and physiological, which will be discussed in the following pages.

In addition, however, this book represents an enquiry into two inter-related problems of desert biology. First, does the extreme variability of the desert climate engender a corresponding diversity of plants and animals? Secondly, are desert plants and animals 'special' in the sense that they have evolved unique mechanisms not to be found in the flora and fauna of other terrestrial environments? Before alluding to these questions in detail, however, attention will be directed to the physical and biotic conditions that are characteristic of deserts, and the ways in which plants and animals have become adapted to them.

The study of ecology discloses underlying principles which enable predictions to be made about the flora and fauna of an environment. However deserts may be defined, the dominating feature common to them all is aridity. The question therefore arises, can aridity alone explain all aspects of desert ecology? One possibility is that only primary producers are limited by water shortage, and that all other trophic levels are determined by that of the primary producers. Another is that water shortage may only affect individuals and have no further direct influence upon the interactions that determine higher levels of ecological organizations, such as populations or ecosystems. In the following chapters, however, it will be shown that desert ecology does not merely encompass the ecology of plants and animals that are adapted to arid conditions. It is the ecology of organisms living in unpredictable, patchy, and variable surroundings that alter both spatially and with the passage of time. The challenge of the desert is not so much in coping with scarcity as with exploiting fleeting opportunities (Safriel *et al.*, 1989).

Desert climates

The physical conditions of the desert environment and its climate have been mentioned so often in the past that it seems unnecessary to discuss them yet again in any great detail (see relevant chapters in McGinnies, Goldman, & Paylore, 1968). Moreover, as Louw & Seely (1982) emphasise, most plants and animals that survive in the desert do so because they, in some way, either evade or avoid true desert conditions (Cloudsley-Thompson & Chadwick, 1964). As far as the fauna is concerned, Louw & Seely (1982) distinguish between escape, a long-term phenomenon, and retreat, a short term behavioural response.

Many short-lived plants and animals evade the extremes of temperature and drought which characterize desert regions. Their survival depends upon adaptations which ensure that they do not experience dehydration in the active state but pass the hot dry season in a resistant passive stage of their life cycle. Perennial desert organisms, on the other hand, are usually adapted to their environment through the possession of avoidance mechanisms rather than the ability of their tissues to tolerate desiccation and very high temperatures. Tolerance usually plays a rather secondary role.

The major deserts of the world result primarily from global, or at least from hemispheric wind patterns, and arise from the whole general circulation. Although many aspects of meteorology are pertinent to arid regions, few of them are unique to deserts of the world. These are often classified from a geographical point of view (Cloudsley-Thompson, 1977) and, more recently, according to the climatic characters that principally affect their flora and fauna (Cloudsley-Thompson 1991).

Arid regions can be divided into five types on a climatic basis. These are as follows: tropical and sub-tropical deserts, cool coastal, rain-shadow, interior continental and polar deserts. In polar deserts, water is not available to plants and animals because it is frozen: in other types of desert, those considered in the present volume, water is deficient throughout most of the year and evaporation greatly exceeds the annual precipitation. Not only is rainfall low and irregular, but low relative humidity has, in itself, an adverse effect upon living organisms because the saturation deficiency of the atmosphere is so great at high temperatures. McGinnies *et al.* (1968), cited above, classified the world's deserts according to mean annual precipitation, as follows:- (a) Extremely arid (less than 60-100 mm), (b) Arid (from 60-100 to 150-250 mm), (c) Semi-arid (from 150-250 to 250-500 mm).

The season during which the rain falls, the duration and amount of precipitation are probably even more important to annual desert plants than to animals. Some 10 mm may be necessary to engender the germination of seeds, even when the temperature and relative humidity are favourable. This varies, however, because water tends to collect on flat plains, in wadi beds etc. whereas it runs off hill slopes.

Whether summer rains, which ameliorate the extreme heat of the summer, or winter rains, which evaporate less rapidly, are more favourable to the fauna and flora is by no means clear. Winter rain is probably more beneficial in general, because it has more time to percolate into the soil. Cool, coastal deserts, such as the Namib and Atacama, seldom receive appreciable rainfall. Fog and mist, however, represent an extremely important alternative source of moisture.

The varying responses of the vegetation to different rainfall regimes is largely based on the monsoon in which nutrients and water are stored for use during prolonged dry periods. Succulents tend to flourish where rainfall occurs in two distant seasons of the year or where fog is a regular phenomenon. Examples are afforded by the rich flora of Cactaceae in the Sonoran desert and of Euphorbiaceae in the Red Sea Hills. In regions with less frequent rainfall, succulents are less common because their internal reserves of water are limited. Instead, ephemerals tend to dominate: the store of nutrients in dry seeds may persist for decades until there is sufficient rainfall to engender germination.

Like average figures for rainfall, mean temperatures are of little significance in arid regions where ambient temperatures oscillate greatly between day and night, summer and winter. The mean maximum temperature in summer may reach 43.6°C in Baghdad, 42.2° C in Bilma, and 40.0°C in Phoenix. Maximum shade temperatures, under standard meteriological conditions, of 58° C have been recorded at E1 Azizia, Libya, and at San Luiz Potosi, Mexico, and 56.5°C in Death Valley, California. The highest ground surface temperature recorded is 84°C at Wadi Halfa, Sudan (Coudsley-Thompson, 1977). Hot dry winds, frequently associated with sand storms, are a constant feature of desert climates.

From the viewpoint of their biota, the world's deserts are, perhaps, best classified as follows:- (a) deserts with seasonal rains (including deserts with winter-mediterranean and summer- monsoon types of rainfall); (b) fog and mist deserts (including cool, coastal deserts where fog appears at irregular intervals through the year); (c) deserts with irregular rain (including some interior continental desert regions and the Australian desert) (Cloudsley-Thompson, 1991).

The Sahara and Arabian deserts are extremely hot and arid, with sparce vegetation cover. This consists predominantly of ephermerals with small shrubs appearing in areas where water collects. Halophytes occur in saline areas, but succulents and grasses are either poorly represented or entirely absent. The deserts of Central Asia are cooler. The Iranian and Turkestan deserts have predominantly winter precipitation, and support stands of small shrubs with *Artemisia herba-alba* being dominant locally. The Thar or Indian desert enjoys summer rainfall and warm winters. The vegetation consists primarily of deciduous shrubs, which lose their leaves during the dry season, and grasses, with halophytes in extensive saline areas.

Of the Great American desert, the Great Basin desert occurs at higher latitudes than the other deserts and is often classified as a cold desert. All four have continental climates. The Great Basin and Mojave deserts have largely winter precipitation while that of the Chihuahuan desert is mostly in summer. The Sonoran desert, located between two seasonal sources of rainfall has a bimodal regime. Evergreen shrubs are dominant throughout: *Artemisia tridentata* on the Great Basin desert, *Larrea tridentata* in the other three. The Great Basin, Mojave and Sonoran deserts have a winter-spring ephemeral flora whereas the Chihuahuan and, to a large extent, the Sonoran also, possess a well developed summer ephemeral flora.

The coastal Atacama-Peruvian and Namib-Kalahari deserts have a dominant flora of lichens with vascular plants, such as *Welwitschia mirabilis* in the Namib, being largely restricted to water-courses. Finally, the Australian desert is characterized by perennial tussock grasses, with *Acacia* trees and shrubs growing along ephemeral water courses, *Atriplex* spp. and other chenopod shrubs in internal drainage basins (McGinnies *et al.*, 1968).

A basic problem facing all desert organisms is the maintenance of an equable temperature without using an excessive amount of water for evaporative cooling. To smaller animals, such as arthropods and reptiles (whose surface to volume ratio is correspondingly high), this problem-always present on land-becomes especially acute. It is significant that the integuments of desert arachnids and insects, and the cuticles of desert plants, both have extremely impervious wax layers, while insect spiracles and plant stomata are especially efficient among desert species (Hadley, 1972). Physiological adaptations such as these are, however, probably of less importance to animals, which are able to retreat into sheltered microhabitats during the heat of the day, than they are to perennial desert plants. Such matters will be discussed in Chapter III.

Despite their necessity for the survival of small animals, desert microclimates have been investigated to a surprisingly small extent. The available information has been reviewed recently (Cloudsley-Thompson, 1991) and can be summarized by saying that, as little as 20 cm below the soil surface, the diurnal range of temperatures is so much reduced that it leads to a daily thermal inversion. Conditions are relatively stable within the burrows of small desert animals -arthropods, reptiles and mammals- so that they do not normally ever experience the full force of the heat and dryness of the desert at midday. Sand and rock provide unique thermal niches in hot deserts and, on the whole, rock temperatures fluctuate even less than sand. Little information is available regarding the protection afforded to animals by vegetation but, given a very small reduction in conditions of extreme heat or drought, an ecological web can depend upon such apparently unrewarding material as dry plant material.

Diversity of growth form among desert plants

Unlike the desert fauna, desert vegetation is often conspicuously different from that found elsewhere. Some of these differences include a preponderance of leafless, stemsucculent shrubs with photosynthetic stems, and an abundance of annual species. Furthermore, deserts are often rich in the variety of growth forms they support, but the reason for this diversity is not easily determined (Cody, 1989). It is, however, probably related to the fact that, unlike desert animals, long-lived desert plants are unable to seek out favourable microhabitats during the day, and are therefore compelled to withstand the full brunt of the desert heat and low atmospheric humidity.

Climate varies widely within the Great American desert region from the Great Basin in the north to the Mojave and Sonoran deserts in the south. In virtually every category of growth forms, including annuals, phanaerophytes and geophytes, diversity also increases from north to south. In phanaerophytes - plants with perennial organs above ground and little or no dieback in the off-season - there is, for instance, a wide range of evergreen, deciduous, nearly or completely leafless trees and shrubs. These may have succulent or non-succulent leaves or stems, of which the latter can, in turn, be photosynthetic or otherwise. In addition, there may be epiphytes, parasites or hemi-parasites. In each of these categories, diversity increases from the Great Basin desert, through the Mojave, to the Sonoran desert (Cody, 1989). Comparable results have been obtained among lichens by Rundel (1978), who showed that the total numbers of species increase at least four times from north to south, while the diversity of fructose, foliose and crustose-squamose growth forms doubles from the Mojave to the coastal fog desert of Baja California.

The situation is similar in the arid interior of Australia. In the south-west, where the climate is somewhat similar to that of the Mojave desert, open bushland is dominated by 'mulga' (*Acacia aneura* and other species of *Acacia*) together with 'spinifex' or porcupine grasses (*Plectrachne* spp.), woody shrubs, and many annual plants. Here the dominant species, *A. aneura*, shows increasing diversity of growth form with decreasing precipitation and higher temperatures (Cody, 1989). In more humid regions, the dominant vegetation type is 'mallee'. In 'mulgas', most of the leaves point upwards so that the rain dribbles down the branches and soaks into the ground beside the trunks. This must surely be adaptive. The species of *Eucalyptus* that dominate 'mallee', on the other hand, present the branching aspect of a pollarded willow. None of these types of vegetation is exclusive, however. 'Spinifex' may grow beneath 'mulga', and saltbush (*Atriplex* spp.), bluebush (*Maireana* spp.), or mitchell grass (*Astrebla* spp.) beneath 'mallee'.

Soil nutrients are in extremely low concentration in this apparently climactic vegetation. The Australian arid zone supports a great diversity of soils, often over relatively short distance, and these add to the diversity of the vegetation growing on them.

Are desert plants and animals special?

According to Louw (1990) the question, 'are desert animals special?' was first posed by the present writer in 1964. Since then, it has had several proponents and, in 1987, a whole conference was devoted to the theme 'What's special about desert ecology?' In summing up that conference, Slobodkin (1989) listed both his initial preconceptions and his subsequent correction of them.

Preconception included: (a) Deserts are physically fragile and the ground does not recover from misuse like the soil of mesic landscapes. Slobodkin amended this by pointing out that, on the largest scale, there are global shifts in temperature and climatic zones which, on a temporal scale of a few years, are barely perceptable. Yet relatively slight changes in global climate may affect rainfall sufficiently to enrich the growth of plants in some deserts, and to reduce it in others. Lack of a protecting organic layer results in a very resistant equilibrium condition. When rain, flood and wind have moved everything that is movable there remains a desert pavement of larger gravel and cobbles overlying and protecting the deeper soils.

(b) Despite their ruggedness in the face of extremes and uncertainties of heat, wind and drought, desert organisms, at least in some cases, are ecologically fragile. This preconception was corrected, in Slobodkin's opinion, by the thought that each particular aspect of the fragility of deserts can be explained by 'a more or less separate mechanism, characterized by its own level of operation and spatio-temporal scale'. The precise solutions to ecological problems therefore differ with the type of organism and with the nature of its environment.

(c) Deserts are simple: for most taxa, fewer species are likely to be encountered in a desert than in a mesic region. This truism obscures the hypothesis of Oksanen & Oksanen (1989) that, once productivity becomes high enough to permit one higher trophic level, there must occur a release from predation on the next trophic layer but one. That is, if there are enough plants to support a population of herbivores sufficiently large that a level of carnivores becomes possible, the latter would be expected to deplete the herbivores to such an extent that the vegetation would become considerably more abundant. A second- level carnivore might, however, reverse that. A low productivity, based on the scarcity of water would, therefore, by itself result in a depauperate set of species.

A third mechanism lowering the number of desert species is an evolutionary one - deserts tend to be ephemeral. Like lakes, they are temporary features of the landscape. The Miocene deserts of Australia may be the oldest in existence, but even they are not older than Lake Tanganyika or Lake Baikal. The present locations of most deserts have not always been desert. Indeed, on a geological time scale, they have not been deserts for very long. This is due, not only to changes in levels of rainfall associated with periods of glaciation, but also to continental drift which has shifted the land masses and the persistent winds that blow over them. 'Is it possible', asked Slobodkin,' to imagine the slow progression of the deserts driving most organisms before it, but permitting those that were pre-adapted to highly variable environments to enter the leading edge, leaving behind those species that had not completely lost their competitive capacity in mesic environments.'

Many terrestrial environments are so homogeneous that plants do not separate into separate niches. This may apply to desert annuals, but perennial desert plants appear always to be segregated into different ecological niches reflected by different growth forms. The rigorous desert environment engenders a wide variety of adaptations and responses in perennial plants because they are immobile, but a somewhat narrower diversity of adaptations is seen in the desert fauna and many of these are behavioural. We shall return to these points later in the book.

2

EVASION OF CLIMATIC EXTREMES

Evasion is the most common adaptation of both plants and animals to the desert environment. It has long been known that the metabolic rate of seeds is so low that they are in a state of suspended animation as, indeed, are arthropods when in diapause. it is not surprising, therefore, that many desert plants should be ephemeral and spend most of their existence in an arrested embryonic stage which is extremely resistant to desiccation and high temperature. When rain falls, however, and there is a brief return of favourable environmental conditions, thesc seeds germinate rapidly, grow to matrutiy, flower, and produce new seeds within the span of a few weeks. Indeed, the African *Boerhaavia repens* is said to take only 8-10 days between sprouting and seed production. Such brief life-cycles are related not only to the restricted duration of the season when conditions are suitable for growth, but also to the shortness of the period during which insects are available for pollination (Cloudsley- Thompson & Chadwick, 1964).

Annual plants

The majority of ephemeral plants are annuals because perennial ephemerals need to maintain more of their dormant tissue—roots, bulbs or corms – during prolonged periods of drought. This is reflected in the preponderance of annuals (96% of plant cover) in Death Valley and other very arid regions of North America where the coefficient of rainfall is about 50%. Where the vairability of precipitation is lower – 20% – the percentage of annuals may drop to as little as 1% of the total plant cover. A similar distribution of annual, versus perennial, ephemerals probably holds good for most of the arid and semi-arid regions of the world, although it is not applicable in certain specialized situations (Louw & Seely, 1982). Because ephemeral plants complete their life-cycles under favourable conditions, they seldom possess morphological adaptations to the desert climate; but several of them employ the C_4 photosynthetic pathway (Chapter IV) which engenders very rapid photosynthesis and growth.

Germination of the seeds of desert annuals does not depend entirely upon a single precipitation, because it is often inhibited by various physiological mechanisms which ensure that all the seeds in a particular area do not germinate at the same time.

These include the presence of inhibitory chemicals in the fruit wall, seed coat, embryo or endosperm. The influence of such organic substances may be reinforced by an osmotic effect. Not only does this complexity introduce considerable diversity in the appearance of germination, but the germination inhibitors are themselves influenced during development.

The fate of the subsequent generation also depends to some extent upon the conditions under which the seeds matured on their mother plant. These conditions vary, in turn, according to the place in which the mother plant was growing, the position of the seeds on the mother plant, the position of the seeds within the pod and so on! In some plants the seeds germinate opportunistically when the chances of survival are low, in others they only germinate when the risk is low– that is, after heavy rain. In some species, mechanisms of both germination and dispersal combine to regulate the number of seeds produced and the time of their germination.

Whereas the genotype influences the fitness of a plant, phenotypic influences during maturation of the seeds ensure that they do not all germinate simultaneously, even when optimal conditions obtain. Phenotypic influences result in an increase in the diversity of seeds, ensuring that only a portion of the seed bank of a species germinates at any one time.

In the Middle East, the seeds of ephemeral plants such as *Zygophyllum* and *Helianthemum* spp. mature and are dispersed by rain at the beginning of summer while those of *Mesembryanthemum, Aizoon* and *Blepharis* spp. are dispersed the following winter. Some, such as the seeds of *Anastatica hierochuntica, Plantago coronopus,* and *Reboudia* spp. are covered with a mucilaginous layer so that they adhere to the soil and are not swept away by floods when rain falls. This phenomenon is known as myxospermy (Gutterman; 1983).

*A. hierochuntica**, the Rose of Jericho, has a mechanism, known as hygrochasy, which limits dispersal of the seeds. As the fruits mature with the onset of the dry season, the leaves fall off and the dry branches roll inwards forming a wickerwork ball. This protects the pods until they uncurl hygrochastically with the onset of rain (Fig.1). The plant has little hold on the soil and, when it is rolled up, is easily detached by high winds and may be blown for a considerable distance.

Hygrochasy is an adaptation restricted to the flora of the desert, and hygrochastic plants are not found in mesic environments. Another rolling plant of the Middle East is *Anteriscus pygmaeus* while, in South Africa, species of *Brunsvegia* have inflorescences that are likewise the sport of the wind, as are those of certain 'spinifex' grasses (*Triodia* and *Plectrachne* spp). in the Australian deserts (Cloudsley-Thompson & Chadwick, 1964)

The fruits of *Colocynthus vulgaris* are spherical, thin-walled gourds about the size of an orange. They are blown about the arid and semi-arid regions of the Sahara, sometimes for considerable distances until they become trapped in a hollow or against a rock. Here they eventually disintegrate or become buried in wind-blown sand. When rain falls and water collects in the hollows, the seeds germinate in favourable conditions.

The caryopses or one-seeded fruits of annual grasses, such as the palaearctic *Aristida funiculata,* are surrounded by two persistent

---

**Anastatica hierochuntica* is known as *Kaff Maryam* throughout the Arab world. The dry plant is reputed for its medicinal value in folk medicine. It is sold in *souks* throughout the Arabian peninsula, and is used by women as a charm at childbirth. The plant is soaked in water and, when it has unfurled, the water is drunk by the expectant mother (Batanouny, 1981)

Fig. 1. *Anastatica hierochuntica* : above, dry; below, wet.

bracts, a lemma and a palea. The lemma is extended at its apex into a long awn which ends in three branches (Fig.2). At the opposite end of the dispersal unit is a group of short stiff horns. One method of dispersal results from the awns becoming entangled, and a large number of dispersal units forming a dense rounded ball which is blown over the surface of the desert by wind. As the sharp pointed ends come into contact with the soil, they anchor the individual units. Eventually the caryopsis disarticulates from the awn and remains partly buried in the soil where it germinates.

The awns and hairs also provide a means of attachment to the coats of animals and are dispersed in this way. The legs of goats sometimes become completely covered with the dispersal units of

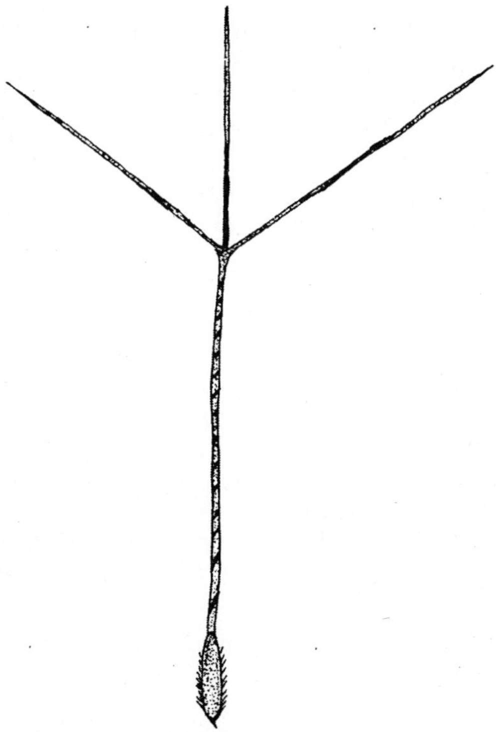

Fig. 2. *Aristida funiculata* dispersal unit.

annual grasses at certain seasons. Many other plants of desert regions have well-developed barbs and bristles which aid their dispersal by animals. Examples are afforded by the spikelets found in bur grasses of the genus *Cenchrus,* as well as in species of *Setaria* and *Tragus.* Some 85% of the seeds of desert annuals are topochorous and are dispersed over short distances: telochoric, or long range dispersal mechanisms are found in only about 15% of the annual plants of arid regions.

Aquatic Arthropoda and diapause

The life cycles of many desert invertebrates show a remarkable parallel to those of ephemeral desert plants because most of the year is spent in a state of arrested development or diapause. During diapause, metabolic demands are minimal, water loss is extremely low, and resistance to high temperatures is marked. Natural depressions may remain dry and apparently lifeless for years yet, within a day or two of being filled with rain water, are occupied by large populations of bacteria, algae and protozoans, which had previously been encysted in the soil. The eggs of phyllopod crustaceans – Anostraca (Fig. 3a), Notostraca e.g. *Triops* spp. (Fig.3b), and Conchostraca (Fig. 3c) hatch almost immediately, and develop

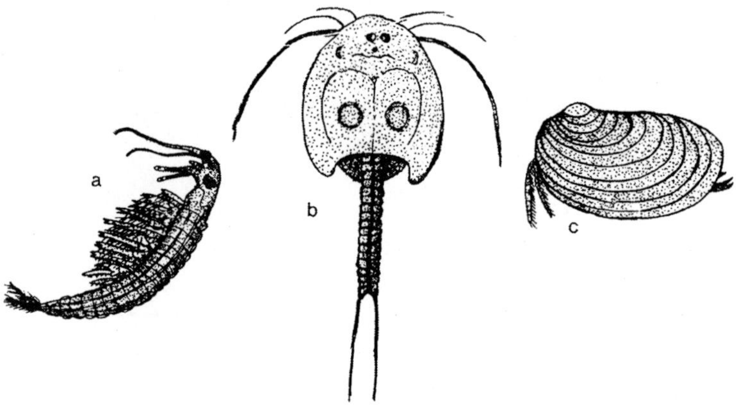

Fig. 3. Phyllopod crustaceans of desert pools: (a) Anostraca; (b) Notostraca; (c) Conchostraca. Not to scale.

extremely rapidly. These crustaceans are the predominant metazoan inhabitants of most desert ponds. Rapid development is essential because the intense insolation causes rapid evaporation, high temperatures and conseqently a low oxygen tension in the water (Cloudsley–Thompson, 1991; Crawford, 1981; Louw & Seely, 1982).

Ephemeral life in temporary pools is not restricted to deserts, but periods of dormancy are longer and wet conditions usually shorter than in other environments. In their review of the adaptational biology of the inhabitants of desert ponds, Belk & Cole (1975) distinguished between obligatory diapause, which occurs every generation and is controlled endogenously, and facultative diapause which is evoked by environmental factors and predict a change to unfavourable conditions. The parallel here with seed germination, discussed above, is remarkable. Clearly natural selection has imposed, upon quite dissimilar organisms, the some response to the unpredictable conditions of the desert.

Although, for most desert arthropods, diapause occurs in the egg stage, in at least 20 species of copepods, such as *Metacyclops minutus*, arrested development takes place in the copepodid stages; while some harpacticoid copepods encyst as adults. The stage in which diapause occurs may well depend upon body size. Animals which are small when full grown would be unable to store sufficient reserves of nutrients, as adults, to sustain life for prolonged periods even though metabolism was greatly reduced. In the embryonic stage, on the other hand, metabolic demands are much lower and less specific (Louw & Seely, 1982).

The function of diapause, like that of seed germination inhibition, is that it ensures that all the individuals of a population do not become active simultaneously. If all the eggs of *Triops granarius* were to hatch the moment they were wetted, for example, a brief shower that did not provide a pool deep enough to last through their entire development cycle would prove fatal to the population.

Aestivation

Like hibernation, aestivation is a period of prolonged dormancy which, in the case of desert arthropods may involve either diapause or quiescence. Consequently, a vernal rain fauna appears at the time of inflorescence when the desert is transformed by an abundance of plant and animal life. Flowers are visited by Lepidoptera, Coleoptera, Hymenoptera and Diptera; termites extend their subterranean galleries to the soil surface and indulge in nuptial flights, whilst predatory insects and arachnids find an abundance of food (Cloudsley-Thompson, 1991). Even species that are active throughout the year usually show seasonal peaks in population density. The desert climate, therefore, by enforcing evasive tactics upon the life-cycles of a wide variety of indigenous plants and animals, imposes uniformity upon a great diversity of organisms.

Many desert invertebrates have annual generations. Examples of univoltine arthropods include woodlice (Isopoda), Solifugae, and many spiders and insects. Among the latter we find numerous Orthoptera, Hemiptera, Neuroptera and Coleoptera. In the case of Lepidoptera, Hymenoptera and Diptera, as Crawford (1981) points out, the distinction between univoltine and multivoltine life cycles is often far from clear. Multivoltine species, with more than one annual generation, include free-living Protozoa and Nematoda, as well as Rotifera and other pseudocoelomates, Acari, and Collembola.

A few female mosquitoes of the species *Anopheles pharaoensis* are able to survive through the long dry season of the Sahel savanna. They take frequent, but incomplete, blood meals which result in a failure of ovarian development. This partial diapause is engendered by the onset of cooler weather accompanied by low humidity. The ovaries undergo only one gonotropic cycle during the dry season and develop extremely slowly so that, when the rains come, the gravid females are ready to oviposit (Cloudsley-Thompson, 1977).

In general, however, dry season diapause is engendered neither by lack of rain nor by a decrease in relative humidity (Denlinzer, 1986). Diapause and quiescence may or may not occur in developmental stages. Either way, a rich fauna of insects appears at the time of inflorescence, while even long-lived animals tend to be much more active then than at other seasons of the year. Despite considerable variability in patterns of reproduction and development, many desert species have life cycles that are correlated with the brief pulses of primary production in deserts. This constraint limits the duration of time during which plants, directly or indirectly, supply their hosts with food and shelter; and simultaneously enhances the value of appropriate timing among consumers (Crawford, 1981).

3

AVOIDANCE OF AND ESCAPE FROM CLIMATIC EXTREMES

The majority of perennial desert plants and long-lived animals are adapted to their arid environments in the possession of avoidance or escape mechanisms, rather than by the ability of their tissues to tolerate high temperatures and extremes of desiccation (Chapter IV). As will become apparent, however, the two are frequently combined — both with one another and with tolerance.

There is a relationship between the density of desert vegetation and the amount of precipitation. Trees may be concentrated in wadi beds where the supply of moisture is higher than elsewhere, whereas shrubs and grasses are often spread much more diffusely across the plains. In this way, isolated plants tend to receive rather similar amounts of water wherever they grow, and their water balances are comparable (Cloudsley-Thompson & Chadwick, 1964). At the same time, they often show morphological adaptations that enhance water uptake and reduce transpiration. The morphological adaptations of desert animals may also protect their tissues so that ambient extremes can be avoided.

Plant morphology

The principal characteristics of many desert plants include the development of extensive rooting systems and the reduction of water loss through transpiration. The water content of the tissues is thereby maintained at a relatively high level. In succulents, the roots spread laterally but are seldom more than 3 or 4 cm from the soil surface. Thus, they exploit to the full every shower of rain, even if the water does not penetrate far into the ground (Fig. 4a). In contrast, desert trees and shrubs have extremely deep roots. *Prosopis farcata* roots may extend to the water table at a depth of 15-20 m in the region around the Dead Sea (Fig.4c). Perennial grasses, such as *Aristida* spp. represent a compromise between these two adaptations (Fig.4b). Some of their roots are extensive and shallow, others penetrate more deeply into the soil.

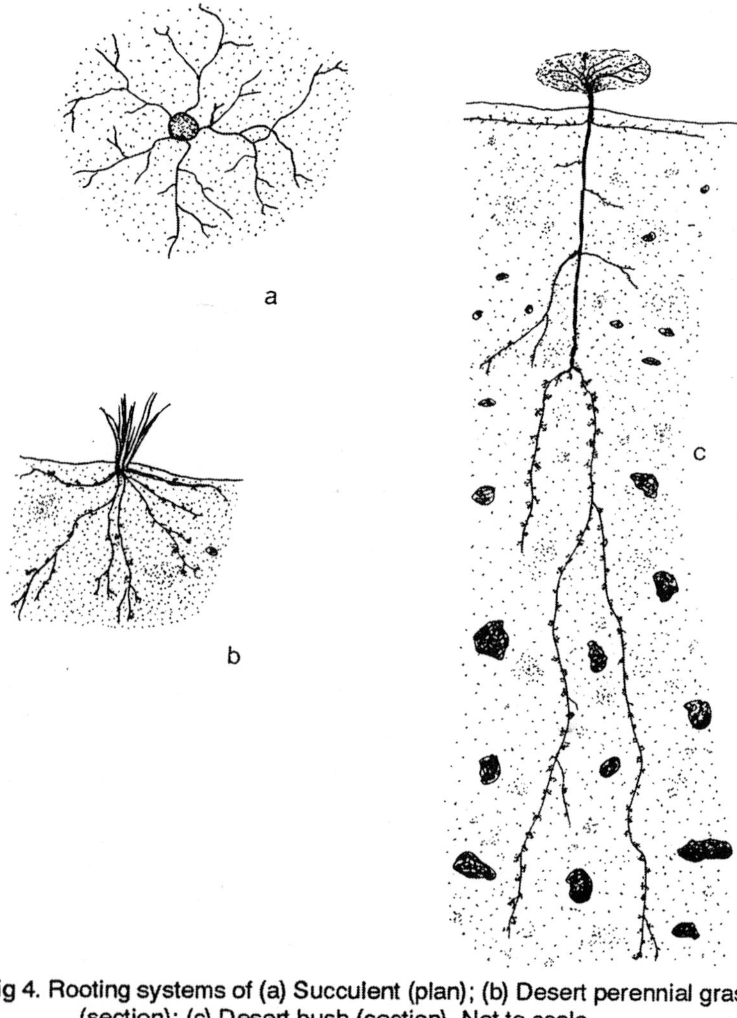

Fig 4. Rooting systems of (a) Succulent (plan); (b) Desert perennial grass (section); (c) Desert bush (section). Not to scale.

The most striking plants in arid regions are the succulents. By storing water in leaves, stalks or roots, these are able to avoid the problem of drought and to survive through long periods of dry weather. Trunk succulents, which have no leaves, include Cactaceae, Euphorbiaceae and Crassulaceae. Cacti occur mainly in America, spurges and sedums in Africa. Most other xeromorphic plants have small, specialized leaves.

A xeromorph is a plant possessing morphological characters which enable it to endure conditions of drought. These include a thick cuticle, reduced leaf size, sunken stomata, small cells, a large amount of mechanical tissue and mucilage cells. Xeromorphy is not restricted to desert plants, however, and by no means all xeromorphs are xerophytes. Thus, not all succulents are xerophytic, although succulence is a xeromorphic character and many succulent Euphorbiaceae and Cactaceae are, indeed, deserticolous.

Cacti showing many strange shapes and forms characterize the desert regions of North, Central and South America. Some are tiny cushions, no more than 5 cm in diameter, others are large spheres or lofty columns reaching upward to the sky. The tallest, *Carnegia gigantea* – the saguaro of the Sonoran desert – may reach a height of 12 m or more, and have a trunk 65 cm in diameter. In advanced stages of growth, it branches out like a candelabrum – for the parallel rising branches, usually five or six in number, only begin to grow when the main stem has reached a height of 5-8 m. The crookedness of the branches results from the fact that they droop, during times of drought and subsequently become erect after rainfall. The growth rate of older specimens averages about one metre per 30 to 50 years, so the age of these giant cacti must reach several hundreds of years and they weigh anything from six to eight tonnes. The side roots of *C. gigantea* may extend up to 30 m in all directions.

Saguaro cacti become reproductive only when they reach a height of several metres. Numerous flowers appear at the apex of the main stem and each branch, and are pollinated by bees of several species. The fruits, 6-9 cm long, are eaten by birds and other animals. They are collected by the Papago Indians and consumed raw or dried. One plant can produce up to 200 fruits in a year each containing about 1000 seeds. These become ripe within a few weeks before the summer rains. An ecological equivalent of this dramatic plant in Africa is *Euphorbia abyssinica,* found in the Red Sea Hills, Sudan.

Quite different cacti are found in the various desert regions of America. In addition, there are other xeromorphic plants such as the yuccas. These have tall trunks bearing a head of rigid leaves with a cloak of dead leaves hanging beside the trunk. The Atacama desert of Chile supports *Tillandsia purpurea* which lives solely upon fog water. The small rosettes of these bromeliads are covered with numerous tiny seeds which take up condensed waters. *Haloxylon* spp., such as the black saxaul (*H. aphyllum*) which thrives on saline clay soils of Central Asia, and the white saxaul (*H. persicum*) of the stony, gypsum deserts of the Kara Kum and Kyzyl Kum, are called 'trees of the desert'. They have bushy stems with leafless branches, can endure extreme drought and tolerate high salinity (Hohn & Petermann, 1980).

The nara (*Acanthosicyos horridus*) of the Namib is a spiny shrub with long sprouts but no leaves. Roots up to 15 m in length and 5 cm in diameter provide the plant with water. *Welwitschia mirabilis* is of interest because it has no close relatives. Individual plants may live to be over 2,000 years old. There are only two leathery leaves which may extend up to four metres in length. These leaves are dissected to such an extent, however, that it is hard to believe that there are only two of them. *W. mirabilis* takes up moisture through its leaves as well as by its roots which extend as much as three metres into the ground.

The orientation of leaves and stems is important to desert plants since it determines not only the surface exposed to light for photosynthesis, but also the heat load obtained from solar radiation. Likewise, vertically orientated pads and stems of cacti absorb less solar energy and remain cooler than those that are orientated at right angles to the sun's rays. It has also been found that pads facing from east to west gain heat more quickly than those oriented in a north-south direction (Louw & Seely, 1982). There is a parallel here with the 'compass' nests of termites (*Amitermes* spp.) in the

Australian desert. Compass nests are also orientated in a north-south direction so that the faces of the mounds are presented towards the sun in the early morning and evening. This warms the nest at times when the atmospheric temperature is low while, at midday, a relatively small area of the termite mound faces the sun.

Surface projections and irregularly spaced surfaces are important in reducing radiation and, therefore, the heat loads on desert plants (Hadley, 1972). In cacti such as cholla *(Opuntia bigelovia)*, the temperature underneath the spines can be as much as 11° C cooler than it would be if there were no spines. This is because spines not only absorb and reflect incident radiation, but they create a boundary layer between the surface of the plant and the tips of the spines which retards heat transfer. Spines also help to protect plants from herbivorous animals (Chapter V).

To conclude – drought evaders (Chapter II), mostly ephemerals but also some perennials, show little obvious specialization for desert living, apart from a high rate of photosynthesis during the brief periods when water is available. Drought avoiders can be divided into three groups (a) Phreatophytes, that is plants whose immensely long roots extend down to the water table. These do not show any other specialization for desert life. (b) Evergreen shrubs which are either leafless or have small and specialized leaves. Their metabolism will be discussed in the following Chapter since they also tolerate heat and drought. (c) Xeromorphic succulents. These are usually leafless, store water in their tissues and possess other physiological adaptations that serve to reduce water loss by transpiration and which are paralleled in desert arthropods (Solbrig & Orians, 1977).

Long term avoidance of drought by invertebrates

The avoidance mechanisms by which the soil fauna survives an environment that is subject to moisture stress for considerable periods of the year fall into three categories: reproductive, behavioral, and physiological (Wallwork, 1982). Among these, there is often no clear distinction between evasion and avoidance.

Unicellular protozoans are essentially aquatic animals yet, paradoxically, some of these microscopic creatures are able to exist in dry sand by the secretion of resistant cyst walls from which their possessors emerge only when the soil has been moistened. Encystment takes place again as the sand dries (Fig.5). In culture solutions, Protozoa normally die or encyst in the temperature range of 44-54° C : encysted forms have lethal temperatures of 70-72° C.

Soil-dwelling nematodes are usually concentrated around the rooting systems of desert plants, the region known as the rhizosphere.

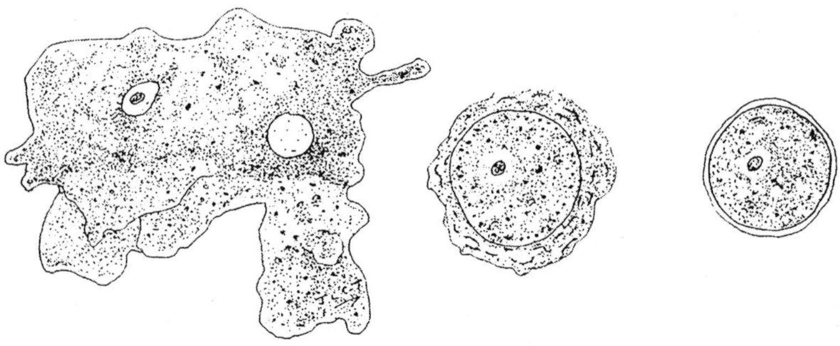

Fig 5. Stages in the encystment of a soil-inhabiting *Amoeba* (length 0.25 mm).

sphere. Many of them, likewise, survive through periods of drought in a quiescent, non-metabolic state of anhydrobiosis which is induced, at least in part, by very slow desiccation. They are not, however, necessarily in a state of diapause, the criterion that I have somewhat arbitrarily evoked to distinguish between evasion and avoidance among desert invertebrates.

The life cycle of the desert woodlouse (*Hemilepistus reaumuri*) provides a well-known example of the timing of reproduction so that this coincides with the season of rainfall. Juvenile instars, which require food and moisture, are recruited into the population when environmental stress is minimal. Other soil arthropods, whose life cycles are correlated with the seasons so that the most susceptible juvenile stages appear at the most favourable period of the year, include pauropods, millipedes, mites and Collembola.

Giant velvet mites (*Dinothrombium* spp.) (Fig.6) emerge briefly onto the desert surface after the rains that trigger simultaneous emergence of the termites upon which they feed. Mating and oviposition take place, and the newly hatched larvae are parasitic upon locusts and grasshoppers. Once the rains have ceased and their termite prey has declined, the mites return to their subterranean burrows where they remain until the following rainy season (Crawford, 1981).

Numerous epigeal arthropods likewise reproduce at the time of rainfall. The remainder of the year is spent in a state of diapause or quiescence. Dramatic examples are afforded by the inhabitants

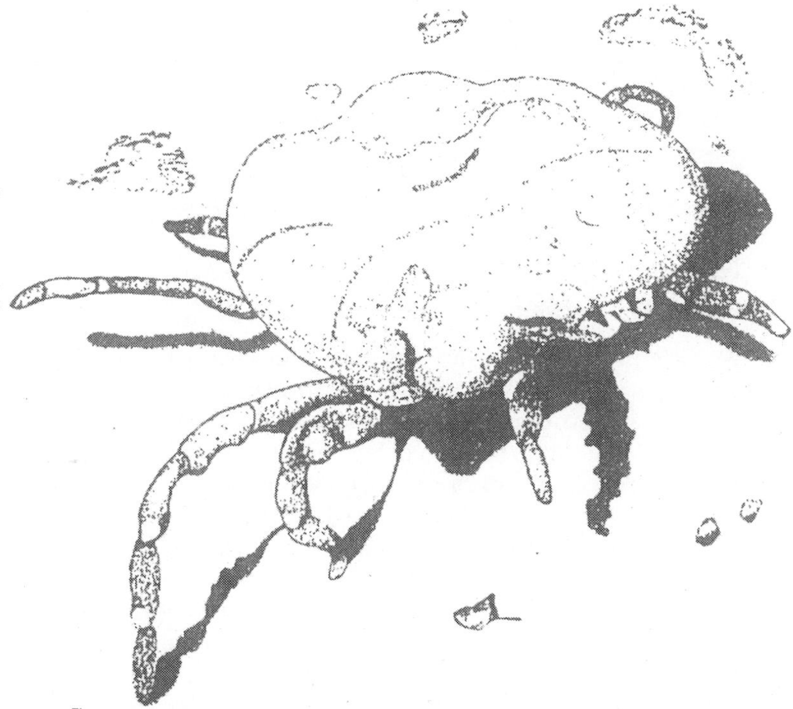

Fig 6. *Dinothrombium* sp. (From Cloudsley-Thompson, 1991)

of temporary rain pools. Within a day or two of their formation, desert ponds and pools become filled not only with bacteria, algae, and protozoans which had previously been encysted, but also, as we have seen, by Anostraca, Notostraca and Conchostraca (Fig. 3) whose diapausing eggs hatch almost immediately they have been wetted and develop extermely rapidly. *Triops granarius*, for example, appears in pools around Khartoum after 7 days and mate within 10: their life span is limited to 20 days (Chapter II). Once again, the distinction between evasion and avoidance is blurred. When diapause occurs, I would regard the phenomenon as evasion but, in the absence of diapause, as avoidance. Arthropods and other invertebrates that pass the dry season in a state of quiescence revive immediately when environmental conditions become favourable. In the case of diapause, there is often no immediate response. I have discussed aestivation in the midge *Polypedilum vanderplancki* (Chironomidae) in Chapter IV as an extreme example of tolerance because, in this and allied species, the tissues themselves dry out and become completely desiccated. There is no diapause there. Diapause and quiescence are physiologically distinct, while the differences between evasion, avoidance and tolerance are mainly semantic. It is far from clear how the classification should be imposed upon phenomena that are often interrelated and grade imperceptibly into one another.

Morphological adaptations of desert animals

Because of their small size and correspondingly large surface to volume ratios, very few terrestrial arthropods are able to cool their bodies by evaporation (Cloudsley-Thompson, 1964, 1991; Edney, 1977). The ability to tolerate high body temperatures (Chapter V), however, is an important factor in the survival of desert insects and arachnids. This is reflected in the low transpiration rates from desert beetles, scorpions and solifugids – a subject that has been investigated in considerable depth. In arthropods, an epicuticular wax layer, as well as the sclerotzed chitin – protein complex of the exocuticle and extensions of lipid material into the underlying cuticular layers, are responsible. Even desert woodlice and millipedes, which lack an epicuticular wax layer, exhibit extremely low rates of water loss through transpiration (Crawford, 1981). A comparable parallel has evolved between arthropods and

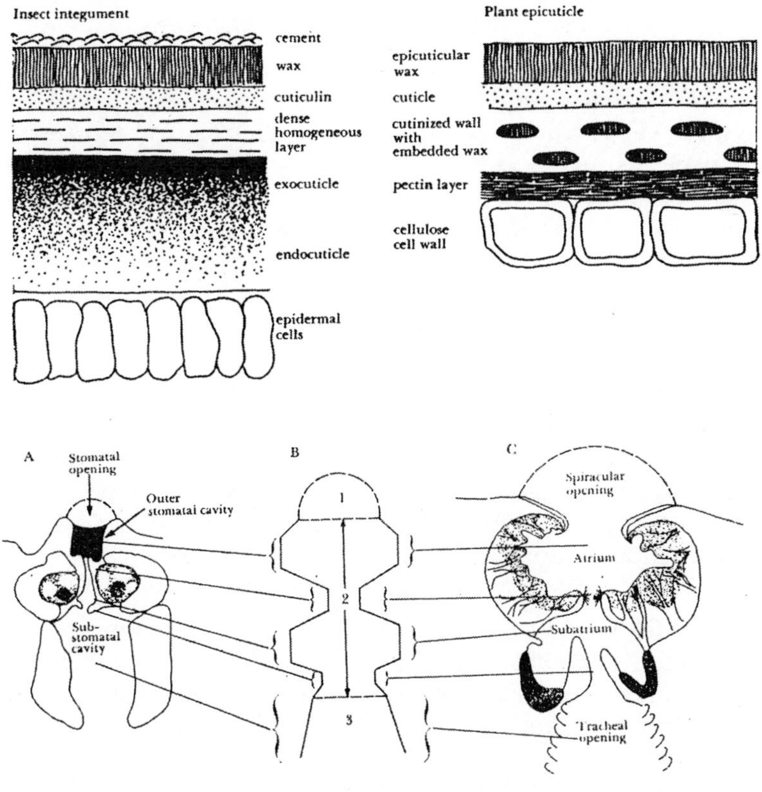

Fig 7. Above: diagrammatic structure of the insect integument compared with that of a plant cuticle. Below: comparison of drought resistance within and outside a plant stoma and insect spiracle. A. Cross-section of stoma of *Yucca*; B. Schematic representation of relative resistance to diffusion through an aperture of varying diameters : 1. External resistance (loundary layer); 2. Stomatal (spiracular) resistance; 3. Sub-stomatal (tracheal) resistance; C. Horizontal section through left spiracle of a desert beetle. (From Cloudsley-Thompson, 1991 from Hadley, 1972).

plants (Fig. 7). The integuments of desert insects and the cuticles of desert plants both have extremely impervious wax layers, while insect spiracles and plant stomata are especially efficient among desert species (Hadley, 1972).

Although few amphibians can be regarded as being true desert animals, evaporative water loss may be considerably reduced in some species during aestivation (see below). Moreover, certain tree frogs such as *Chiromantis xerampelina* of southern Africa are able to survive in hot savanna without access to water by reducing evaporative water loss throughout the dry season. This reduction in the permeability of the integument is due to the presence of a dense layer of chromatophores in the skin of the back. In the case of South American *Phyllomedusa* spp., however, low permeability is conferred by a lipid film secreted on the surface o the skin.

Water loss through the integument of reptiles has been studied in great depth (Cloudsley-Thompson, 1971, 1991) and will not, therefore, be discussed at length here. In general, it may be concluded that, as in arthropods, control of water loss is primarily physiological or morphological in desert reptiles whereas, in contrast, thermoregulation is achieved primarily by behavioral means.

Bird feathers not only act as a shield against excessive insolation in the desert but also facilitate loss of heat through convection and radiation. A satisfactory compromise between these two conflicting functions is achieved in the plumage of the ostrich (*Struthio camelus*) whose feathers are both long and sparse. At high ambient temperatures, the ostrich erects the feathers on its back, thereby increasing the thickness of the barrier between solar radiation and the shin. At the same time, this facilitates an increase in the lateral movement of air over the skin. This adaptation is seen to a less extent in smaller carinate birds which are able to fly into shade or soar high above the hot desert surface. Nevertheless it is to be found among sandgrouse (*Pterocles* spp.) which nest on the ground and need to shade their eggs during the day (Louw & Seely, 1982).

The pelages of desert mammals face similar conflicting demands. The coarse hair of the camel, for example, not only acts as a barrier to solar radiation and slows the conduction of heat from the environment to the body of the animal, but it allows sweat to evaporate on its skin so that maximum evaporative cooling is obtained. Much of the advantage of sweating is lost to horses, whose hair is so dense that the sweat has to pass through it before it evaporates. The camel stores reserve fat in its hump so that the remainder of the body surface can function as a radiator. The same is true of Indian and Zebu cattle while, in desert sheep, fat is stored in the tail.

The reader will have noticed that, both in arthropods and to a lesser extent, in reptiles, a lipid layer covers the entire body. In mammals, on the other hand, fat is not evenly dispersed over the body surface. The difference is related directly to body size. The surface to volume ratio of small animals is so high that they cannot afford to cool their bodies by transpiration. In contrast, large endothermic animals such as camels, with small surface to volume ratios, cannot afford not to! Of course, camels only sweat as a last resort. They allow their temperatures to vary over a greater range than most other mammals do and do not begin to sweat until their bodies have reached 40.7° C. Heat is stored during the day and dissipated at night without expence of water. Moreover, when the difference in temperature between the hot environment and the animal's body is smaller, less heat reaches the body and less water is required to prevent a further increase in body temperature.

The above discussion illustrates the effect of size upon an otherwise comparatively simple relationship between water conservation and thermoregulation. In this context, the diversity of desert animals primarily reflect their sizes.

Before turning to other matters, I would like just to comment upon the function of sweating. Unless evaporation takes place on the skin – as in camels and human beings – the water expended is largely wasted. Desert gazelles and horses can sweat but, because of their fine hair, this is a most uneconomical use of precious water. It would normally be employed only in an emergency as for example, when escaping from a predator. In contrast, jerboas and kangaroo-rats do not sweat. When their body temperatures rise towards lethal limits, however, they salivate profusely wetting the hairs of their chests. This emergency cooling may keep them alive for 20 minutes or so, which could mean the difference between life and death. Tortoises do the same, wetting their necks and front legs with saliva, and their back legs with urine, when overheated. This, again, is a form of thermoregulation used only in emergencies (Chapter IV).

Aestivation in vertebrates

Because of their dependence upon water, amphibians are restricted to specialized habitats in desert regions. The spadefoot toads (*Scaphiopus* spp.) of the Sonoran desert, for example, dig burrows to a depth of 90 cm or more, where they remain periods as long as 9 months. Burrowing is accomplished with the aid of a horny projection on the hind foot. While aestivating in a state of retraherence, their metabolic rate is greatly reduced and they depend upon reserves in the fat bodies within the abdominal cavity.

Whilst aestivating, some amphibians surround themselves with cocoons of mud, others with dead skin, which is relatively impervious to water loss. Some accumulate urea in their tissues. This raises the osmotic concentration of the body fluid above that of the surrounding soil, so that moisture passes from the surroundings into the animal's body. Retention of urea in the body, however, is not restricted to desert species: it is found in the marine frog *Rana cancrivora*. Survival is also enhanced by storage of water in the bladder and tolerance of marked dehydration. A loss of 40-50 per cent of body weight frequently occurs without having any harmful effects. In contrast, the African toad *Bufo regularis*, which does not aestivate, only breeds in irrigation ditches and seasonal pools where these are just beyond the cultivated areas fringing rivers. It has not become adapted to conserve water by reduced evaporation or an increased water uptake and, in these respects, compares with the common toad (*B. bufo*) of temperate regions. However, it has a lower rate of evaporation that *B. mauritanicus*, which is able to maintain a greater thermal differential with the environment by evaporative cooling. Although it cannot survive for long in a hot, dry atmosphere, this ability to maintain a lower body temperature for short periods in dry air may be useful in hot habitats, possibly explaining its presence in the Sahara (Lambert, 1984).

Many desert reptiles hibernate in winter when ambient temperatures are too cold for them to be active and food is scarce. Some, too, retreat from prolonged heat in the same way, by aestiv-tion. This is practised by the Gila monster (*Heloderma suspectum*) which lives in arid surroundings without being adapted to tolerate high temperatures. Most desert reptiles hibernate or aestivate in burrows or in rock crevices.

Whereas ground squirrels that inhabit the cold northern latitudes of the Nearctic region are able to pass the winter in a state of hibernation, their relatives of tropical and sub-tropical deserts to the south, such as the Mojave desert ground squirrel (*Citellus mohavensis*), spend the dry season in aestivation. Species which inhabit temperate regions neither hibernate nor aestivate, so no sharp distinction should be drawn between the two phenomena. Among homeotherms, aestivation is most common in rodents, but it has also been recorded in the marsupial *Cercaertus nanus*. The best-known example of hibernation in birds is provided by the poorwill (*Phalaenoptilus nuttallii*) of the Californian desert. Not only does this species exhibit long periods of dormancy in the dry season when food is scarce, but its basal metabolism under summer conditions is still less than half that of most other birds (Hudson & Bartholomew, 1964).

Phenology

Even animals that are active throughout the year and neither aestivate nor hibernate usually show seasonal rhythms of activity and especially seasonal reproductive cycles. Seasonal rhythms occur in the diel activity of some species, both arthropods and reptiles, which are nocturnal in summer but are crepuscular or are day-active during the cool season of the year. Furthermore, species that are active throughout the year often show seasonal peaks in population density. These are correlated with a number of factors including rainfall and plant growth as is the phenology of reproduction.

Life history of patterns are as variable among desert animals as they are among the inhabitants of other biomes. Despite this, however, they are more closely related to rainfall and plant growth than they are in less extreme environments in which temperature and photoperiod are relatively more important as synchronizers of reproduction. This applied both to invertebrates and to vertebrates. Among lizards, for example, reproduction is primarily correlated with precipitation and humidity although temperature may also play a part in regulating androgen biosynthesis and is an important environmental cue for reproduction. An endogenous circannual rhythm may also be involved (Cloudsley-Thompson, 1991).

Dromedaries (*Camelus dromedarius*) mate at the time of the annual rains, but the baby is not born until a year later when there is again plenty of food for its mother. In Bactrian camels (*C. bactrianus*), the gestation period is even longer – a period of 370 to 440 days. In both species, the young are born singly and are suckled by their mothers for three or four months at a time when there is plenty of lush vegetation and no lack of moisture.

Intraspecific differences have been observed in the reproduction of mountain gazelles (*Gazella gazella*) in Israel. In the eastern Lower Galilee, where they have free access to water throughout the year, females reproduce during every month of the year, conceive first at the age of six months and exercise long parental care. Females of populations further north in Upper Galilee, on the other hand, maintain a seasonal reproductive pattern, do not conceive until they are 18 months old, and have a short period of parental care. This pattern is similar to that of the dorcas gazelle (*Gazella dorcas*) in the south of the Negev desert. Interspecific similarities between these latter populations are attributed to the lack of free water during lactation, when the gazelles have to rely upon water available in plant tissues only. Apparently the timing of births and the ability of female gazelles to conceive whilst lactating are not species-specific characters but simply a matter of short-term adaptation to changing environmental conditions (Baharov, 1983).

Adaptations of parasites

Desert temperature cycles affect parasites whose life cycle is not confined to the bodies of their hosts, just as they affect other organisms that inhabit arid regions. For instance, the reproduction of the monogenetic trematode *Pseudodiplorchis americanus*, a parasite of the North American spadefoot toad (*Scaphiopus couchii*), is adapted to a strict annual cycle. There is an extremely brief opportunity for transmission, restricted to less than 24 h in the year when the host enters water to breed. This means that invading flukes have a whole year in which to complete their reproductive development, and accumulate their entire reproductive output *in utero* during the time that the toads are in hibernation. Nevertheless, a large proportion of the population of parasites is not prepared for transmission at the time that the toads are breeding. The effective period for accumulation of the infective stages of *P. americanus* in any one season varies according to the timing of the summer rains in consecutive years. Since the first month after infection consists of pre-reproductive development, first-year worms have only 3-4.5 months in which to produce infective larvae. in the shortest seasons, there may be insufficient time in which to complete the reproductive programme in time for the first opportunity for transmission. In different parts of the range of *S. couchii*, there are different cycles of temperature, rainfall and breeding. Varying thermal regimes consequently represent an important abiotic control on the reproductive biology and transmission of *P. americanus* (Tocque & Tinsley, 1991).

Migration

The most obvious way of escaping from the rigours of the dry season in semi-arid regions is by migration; but, of course, this can only be undertaken by mobile animals such as birds, bats, and large mammals – principally antelopes. The present book, however, is not the place in which to discuss such migration, and I shall not elaborate further.

Retreat

Louw & Seely (1982) use the term retreat in the context of the daily rhythm of surface activity and rest, characteristic of the majority of small desert animals especially arthropods, reptiles, and mammals like jerboas, kangaroo-rats, and gerbils. The diurnal locomotory rhythms of such creatures are almost invariably controlled by endogenous circadian clocks which are entrained by fluctuating cycles of light and/or temperature, the relative effects of which vary among different species (Applin *et al.*, 1987).

Not only do many desert animals show a bimodal activity rhythm which allows them to escape the fierce heat of the midday hours, but many desert lizards shuttle between sunshine and shade thereby thermoregulating behaviourally. Rhythmic locomotory activity has been analysed in many desert arthropods and reptiles and the subject has been reviewed recently (Cloudsley-Thompson, 1991). Some confusion has been caused by the fact that, although aktograph experiments in the laboratory invariably show a marked circadian rhythm in scorpions, field observations indicate that, although emergence at dusk of *Paruroctonus mesaensis* in California is correlated with decreasing levels of ambient light intensity, scorpions do not leave their burrows every night (Polis, 1980). The explanation may well be that, although their biological clocks may stimulate the scorpions to come to the mouths of their burrows, they do not actually emerge unless they require food and conditions outside are favourable.

As in arthropods, the daily activities of desert reptiles are controlled by endogenous circadian rhythms. The time of day at which a reptile is active plays an important part in its adaptation to life in the desert. The daily and seasonal patterns of reptilian behaviour are strongly influenced by their thermal biology. In general, day-active arthropods and reptiles tend to maintain high body temperatures, often approaching lethal levels, but shuttle into the shade when in danger of becoming overheated. This is by far their most important means of thermoregulation.

Among endotherms - birds and mammals – the situation is quite different. With few exceptions such as owls and nightjars, birds are day active. Some large desert forms, such as eagles, kites and vultures, are able to soar in the cool air far above the scorching heat of the desert sand. Apart from being generally paler in colour desert species are indistinguishable from their relatives in humid regions. Although many species have been sighted in arid lands, they are seldom found beyond the range of free water unless, of course, they are migrating. Water loss by evaporation is the most serious physiological factor limiting their distribution. Small birds that cannot keep cool by high flying rest in the shade as much as possible during the hottest part of the day. Even so, they are exposed to much more radiant energy than are burrowing rodents.

Owls and nightjars are an exception and hide deep in clefts and fissures of the rocks during the hours of daylight.

Early in the morning, most desert birds feed actively. Later in the day, the smaller species shelter in trees and bushes, or in the shadows of rocks, with their wings hanging loosely and their beaks open as they pant, cooling themselves by respiratory evaporation. The largest bird of all, the ostrich, like camels and antelope, minimises water loss by behavioural thermoregulation, orienting to the sun and moving its wings downwards and away from the thorax so that the naked surface of the thorax is shaded while radiant and convective heat loss are facilitated. Physiological regulation of water loss will be considered in the following Chapter.

Most small mammals are nocturnal, and spend the hours of daylight in their burrows, thus avoiding the most unfavourable extremes of heat and drought. An interesting exception is afforded by the African ground squirrel (*Xerus inauris*). On hot, windless days, this animal raises its expanded tail above its back, thereby shading itself as it forages for food. Numerous other examples of escape from climatic extremes could be cited. The important point, as far as the present book is concerned, is that avoidance of climatic extremes, in one way or another, is by far the most important adaptation of most desert organisms and the same methods are employed by often quite unrelated forms.

4

TOLERANCE OF CLIMATIC EXTREMES

Although most small animals survive in the desert largely through evasive behaviour and avoidance of climatic extemes, a few of them also show adaptations that permit them to some extent to tolerate these extremes. Tolerance, however, is of especial significance to perennial plants that do not evade, or avoid, environmental conditions morphologically, and to larger animals that are unable to avoid the exigencies of the desert climate. Small desert animals, such as arthropods and reptiles, tend to exhibit behavioural avoidance of high temperatures and physiological responses which enable them both to avoid and to tolerate drought. Scarcely any plants and animals, however, are able to withstand complete dehydration of their tissues. Examples of the few that can do so are afforded by poikelohygric cryptograms and certain almost unique insects which are able to withstand complete desiccation. These will be discussed below.

The major environmental adaptations of desert animals include: (a) Size and shape, (b) Orientation to the sun, (c) Colour and nature of the integument, (d) Absorption and storage of water, (e) Facultative reproduction, (f) Tolerance of high temperatures, (g) Tolerance of dehydration, (h) Specialized respiratory patterns, (i) Conservation of water lost in excretion and by evaporation, (j) Reduction in metabolic rate, (k) Modifications in digestive physiology, (l) Escape from or avoidance of unfavourable conditions

(Louw & Seely, 1982). Some of these adaptations have already been discussed, but others enhance tolerance of the desert environment. The latter include the tolerance of desiccation and of high temperatures, thermoregulation, and physiological conservation of water. Reduction in metabolic rate is of significance more in relation to the shortage of food than in thermoregulation. By no means are all these adaptations to be found in every species or taxon, but they are widely spread throughout the animal kingdom and many are also found in plants.

Physiological conservation of water by animals

In the preceding chapter, I pointed out that both arthropods and plants are able, to some extent, to avoid desiccating conditions because they possess outer coverings of impervious wax and respiratory apertures whose structure reduces the amount of water lost by transpiration. Water loss is also reduced among desert animals by the evolution of extremely efficient excretory systems to be discussed below. Before we turn to this subject, however, the conflict between the necessity to conserve water, and the advantages of evaporative cooling among arthropods deserves further attention.

Although the phenomenon of evaporative cooling in insects has been observed in many species, including cockroaches, grasshoppers and locusts, beetles and bees (Edney 1977), it is not common among arthropods, whose capacity to maintain water balance under stressful conditions is due mainly to their ability to excrete insoluble nitrogenous waste products and the low permeability of their cuticles (Chapter III).

An important exception to this generalization is the American desert cicada (*Diceroprocta apache*). By feeding on the xylem of desert plants, this species has access to a constant and abundant supply of water, like the phreatophytic plants discussed earlier (Chapter III). Water is extruded actively through pores on the dorsal surface of the thorax and abdomen, in a manner similar to thermoregulatory sweating in mammals. This release of water begins at the point where cicadas seek milder microclimates, and it enables them to prevent their body temperatures from reaching lethal limits. Simultaneous measurements of water loss and gaseous exchange from cicadas feeding on perfused twigs show substantial increases in transpiration and temperatures at which evaporative cooling begins (37-38°), but only small increases in oxygen consumption and the production of carbon dioxide. The extent and duration of evaporative cooling depend both upon the cicada's state of hydration and the rate of water loss from the abdominal cuticular pores (Hadley *et al.*, 1991). To some extent, therefore, desert cicadas parallel deeply-rooted desert plants which evade the problem of drought in the desert.

In contrast, most desert arthropods are without sources of water and have extremely low rates of transpiration. Particularly remarkable in this respect are Coleoptera, scorpions and Solifugae. The Acari is the only arachnid order in which the active uptake of moisture from unsaturated air has so far been demonstrated but it also occurs in Thysanura flea prepupae, and in females of the Mojave desert cockroach (*Arenivaga investigata*). Whether water is absorbed through the cuticle, the rectum, or both, has not been fully established; nor has the physiological mechanism involved. So far, the uptake of moisture in this way has only been demonstrated in wingless forms: its apparent absence in Tenebrionidae may reflect their extremely low rates of transpiration. On the other hand, there could well be a region of the tenebrionid abdomen that absorbs water vapour from the sub-elytal cavity (Edney, 1977).

Loss of water accompanying the elimination of nitrogenous waste is minimized among insects by the excretion of insoluble uric acid and, in arachnids, of guanine as the endproducts of protein catabolism. The histophysiology of excretion is best explained by reference to Fig. 8. In the first stage, an alkaline solution of potassium and sodium urates is secreted at the distal end of the malpighian tubule. At the proximal end, water is absorbed, the solution becomes slightly acid, and uric acid crystals are precipitated. These pass into the hind gut where more water is absorbed by the rectal gland. Finally, a dry faecal pellet containing crystals of uric acid wrapped in peritrophic membrane is expelled to the exterior. There are numerous elaborations on this basic theme.

The chief excretory organs of arachnids are the coxal glands, but malpighian tubules are also present. These are not homologous

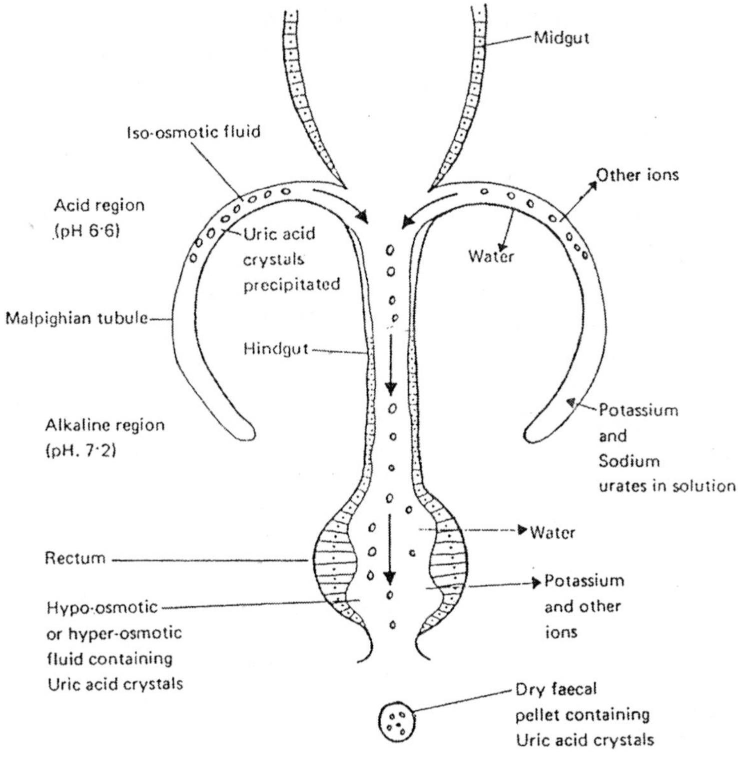

Fig. 8. Excretory processes in an insect (*Rhodnius*). (From Cloudsley-Thompson after V.B. Wigglesworth).

with the malpighian tubules of insects – another example of parallel evolution in different taxa – and several other types of cells and glands may also have an excretory function among arthropods.

Most desert insects and arachnids are unable to tolerate any greater loss of water than are their relatives from more humid places although, as emphasised above, they are more resistant to water loss by evaporation. Whereas the osmolarity of the haemolymph is regulated in desert Tenebrionidae, scorpions can tolerate increasing osmotic pressures when subjected to desiccation. Heat death is associated with a decrease in the pH of the blood, and may be related to accumulation of acid metabolites at high temperatures faster than they can be eliminated (Cloudsley-Thompson, 1991; Edney, 1977).

Amphibians may be divided into three groups according to their susceptibility to water loss. At one extreme are species whose skin is highly permeable to water. At the other, is a group which contains species that are able to tolerate desiccation because their skins are highly impermeable. These have rates of water loss only 1-5 per cent of those found among members of the intermediate group. They include species from several phylogenetically diverse taxa, including the genera *Chiromantis* (Rhacophoridae), *Litoria* and *Phyllomedusa* (Hylidae), and *Hyperolius* (Hyperoliidae). The desiccating conditions of arid environments are enhanced for some species by the fact that they normally thermoregulate by basking in the sun. For instance, the Canyon tree frog (*Hyla arenicola*) basks on vertical rockfaces during the day. This species shows an intermediate rate of evaporative water loss, but it is never found far from permanent water and is therefore able to rehydrate itself on a daily basis. Moreover, its enlarged bladder enables it to store up to 25 per cent of its body mass in dilute urine from which the body fluids can be replenished. Dehydration is therefore tolerated throughout the day because this confers advantages of evaporative cooling. But why these tree frogs should need to bask in the sun during the heat of the day is by no means clear.

Most desert amphibians respond to dehydration by decreasing the rate of urinary water loss and increasing the rate at which water can be absorbed through the skin. This response is mediated by a neurohypophysial hormone, arginine vasotocin. Little urine is produced, and urea is stored in the body when water is restricted. Desert amphibians can also tolerate high concentrations of urea in their blood and tissues while they are aestivating or hibernating.

Both reptiles and birds lay cleidoic eggs and, consequently, are uricotelic. This pre-adapts them for life in arid environments, and further desert adaptation can only be a matter of degree. When partially dehydrated, desert species typically extract a higher percentage of water from their faeces than do their relatives from more humid climates.

The ability to minimize excretory water loss in reptiles is associated with the lack of a loop of Henle in the reptilian nephron. Reptiles, therefore, cannot produce hypertonic urine, although *Amphibolurus maculosus*, a lizard which inhabits dry salt lakes in South Australia, voids a urine that is hyperosmotic to the plasma. Urine is concentrated in the cloaca and not in the kidney as it is among birds and mammals. Most desert reptiles are either partially or completely uricotelic, although urate may not be excreted solely as uric acid. Tortoises, in particular, show considerable temporal variations in the ratio of urea to uric acid excreted and, in several species, changes from a predominantly uric acid to urea metabolism and vice versa may occur within a single individual. Tortoises and other desert reptiles tolerate greater changes in the volumes of their body fluids, and in the osmotic concentrations of their extracellular fluids, than do birds and mammals. In addition, lizards may possess extrarenal excretory organs in the form of nasal salt glands which secrete fluids rich in potassium (Cloudlsey-Thompson, 1991).

Unlike amphibians and reptiles, birds and mammals are able to excrete a fairly concentrated urine. Moreover, the fact that the ureters of birds branch to join the kidney tubules directly – there being no renal pelvis as in mammals – is a key structural feature that enables birds to utilize uricotelism whilst simultaneously producing an osmotically concentrated urine. This is forced by peristalsis from the tubules directly through the branches of the ureters into the cloaca. Even desert species have a rather low tolerance to saline waters, although some can exist on a saline solution slightly stronger than sea water. A few, including the ostrich, possess nasal salt glands.

Jerboas and kangaroo rats can survive indefinitely upon dry food without any water to drink, but they avoid the midday heat by burrowing (Chapter III). They are able to excrete large amounts of salt, and yet maintain normal water balance. They can also excrete load of urea of about 23 per cent. Camels and other larger desert mammals are also able to excrete concentrated urine but, in many of them, the urea content is low (Cloudsley- Thompson, 1977).

Throughout the deserts of the Old World, antelope have acquired the ability to become independent of drinking water. Grant's gazelle (*Gazella granti*) and oryx (*Oryx beisa*) can survive under desert conditions provided that shade is available and there is over 30% moisture in their food. Eland (*Taurotragus oryx*) can also exist without drinking but require access to shade and food containing 60% moisture. Likewise, the Arabian oryx (*Oryx leucoryx*) can do without drinking in the wild at an average maximum temperature of 40° C if there is 50% moisture in their diet while black buck (*Antelope cervicapra*) need about 40% moisture in their food during the winter and need to drink water at least once in 4 to 5 days in summer. The situation is comparable with that of the dorcas gazelle (*Gazella dorcas*) and its ecological equivalent, the springbok (*Antidorcas marsupialis*).

Tolerance of dehydration of the tissues

Although cryptograms – lichens, mosses and lycopods are poorly represented in some deserts, they are quite abundant in others, particularly coastal deserts. Most cryptograms lack stomata and desert forms are poikelohygric. They can tolerate repeated and prolonged dehydration, yet immediately become active upon rehydration and resume photosynthesis. They equilibrate with the level of atmospheric humidity, tolerate extreme heat and cold when desiccated, and take up water directly from the air. Desert ferns are an exception because, although they possess stomata, they are similar to poikilohygric plants in other respects. Poikilohygry is a useful adaptation in coastal deserts where dew is almost the only source of water but, because of their low photosynthetic capacity and brief periods of activity, desert cryptograms can only be successful where higher plants are very scarce or absent (Smith & Nobel, 1986).

The ecological dominance of the creosote bush (*Larrea tridentata*) in the plant communities of the warm deserts of North America is evidence of its adaptation to arid environments. This adaptation takes the form of tolerance of stress. The leaves are capable of passing into a resting phase upon severe drying and, during this period, appear brown and brittle. Stresses that limit productivity are mainly environmental – shortage of water and low temperature – although others such as shading and shortage of nutrients may be due to competition. Certain characteristics of the stress-tolerant strategy, such as leaf size, are structural while others, such as low rates of photosynthesis, are physiological. Since this is a reflection of the small xerophytic leaves, small cells, thick cell walls, and so on, it can be seen that structure and function, once again, are interdependent (Whitford, 1986).

We saw, in Chapter II, how many arthropods are able to evade the prolonged droughts of the desert climate in a state of diapause when their rate of metabolism is much reduced although they are fully hydrated. In contrast, the larvae of the chiromomid midge *Polypedilum vanderplanki* can tolerate almost total dehydration in a state of cryptobiosis and not in diapause. This species inhabits shallow rock pools which may fill and dry several times during the rainy season. Each time that rain falls the larvae absorb water through their integument and begin feeding within an hour. They also tolerate immersion in liquid air (-190°C) or liquid nitrogen (-270°C) yet subsequently produce normal adults. For hydrated larvae, on the other hand, thermal extremes are fatal (Hinton, 1960).

Photosynthesis and crassulacean acid metabolism

The reduction of carbon dioxide and consequent synthesis of carbohydrates by plants under the influence of sunlight is achieved through the Calvin metabolic cycle. In most plants (C_3 plants), carbon dioxide enters the Calvin cycle directly. Certain xerophytes, however, keep their stomata closed during the day and open them at night when the temperature and saturation deficiency of the atmosphere drops. Carbon dioxide is then absorbed and combined with phosphoenolpyruvate to produce malate which accumulates in the cells. When the sun rises, these organic acids are broken down, releasing carbon dioxide which is incorporated into carbohydrates by the enzymes of the Calvin cycle. This pattern of metabolism, known as crassulacean acid metabolism (CAM), is especially common among succulents, including members of the families Crassulaceae, Cactaceae and Bromeliaceae. Plants in which carbon dioxide is fixed only in daylight, are known as C_4

plants. Winter ephemerals are usually C_3, while summer ephemerals, such as grasses, are C_4 plants (Louw & Seely, 1982).

A general set of photosynthetic responses characterizes each structural/functional floral group. Arid regions with winter rainfall and a long summer dry season are relatively rich in evergreen shrubs and winter ephemerals, with few succulents or perennial grasses. In contrast, semi-arid regions with a predictable wet and relatively short dry season are prone to be dominated by deciduous shrubs and perennial grasses. A fundamental trade-off occurs between photosynthetic capacity among leaf-bearing vascular plants, and the ability to remain metabolically active at low water potentials. The dichotomy in photosynthetic performance between C_3 and C_4 plants tends to break down as water stress becomes more important. Consequently, ecophysiological differences between C_3 and C_4 evergreens are much less pronounced than in C_3 versus C_4 annuals or grasses (Smith & Nobel, 1986).

Thermoregulation in animals

Few animals are able to withstand excessive heating of their cells and tissues. Almost without exception, desert animals seem to avoid excessive heat by behavioural means. Only the largest, with very small surface to volume ratios, can, in general, also regulate their body temperatures by evaporative cooling – and even camels avoid insolation whenever possible.

Because ectotherms such as arthropods and reptiles have much lower metabolic rates and higher lethal temperatures than endotherms, they are not subject to the same limitations of body size on their ability to afford to utilize water for evaporation cooling. The desert cicada, discussed above, provides an example. The common assumption that small body mass precludes the adaptation to desert conditions is not, therefore, necessarily invariably valid but, as a generalization, it is correct. Furthermore, because of their large surface to volume ratios, small animals tend to absorb more heat than do large ones when their body temperatures are exceeded by the ambient temperature.

Another important point to remember is that the various physiological systems of thermoregulation within an animal do not operate in isolation, but interact with one another. In the natural environment, different mechanisms may predominate at different times and under different conditions, within the same individual. At the same time, in different species of the same taxon, there may be a variety of different responses depending upon the heat load and the availability of water in different micro-habitats.

The thoracic muscles of insects generate a considerable amount of heat during flight but, in small species, the rate of convective heat loss is so rapid that little of this is retained in the muscles. On the other hand, in larger forms such as locusts, bees and moths, obligatory endothermy may raise thoracic temperatures more than 30°C above the ambient. To maintain this high rate of metabolism, the flight muscles must be provided with a continuous supply of oxygen *via* the tracheae and tracheoles. In consequence, ventilation is accompanied by evaporative water loss which increases as the temperature of the thoracic muscles rises. The higher the temperature, the more evaporation takes place and the greater the amount of cooling. Reduction of temperature also takes place through the transport of heat by the blood from the thorax to the abdomen when there is a large thermal excess. Physiological limitations place great importance on behaviour that ameliorates thermal and water stress. The locust's relatively low metabolic rate in flight, together with nervous control of the spiracles, appears to make this insect well adpated to the desert.

Small reptiles have a higher rate of water loss per unit body weight than do larger ones, the difference being related to the relatively larger surface area in small animals. Panting sets in at higher temperatures, however, and this increases evaporative cooling. Thermoregulatory salivation occurs as an emergency mechanism in tortoises, but it is not restricted to desert species. In *Geochelone sulcata* from the African sahel, for example, evaporative water loss increases greatly when the ambient temperature exceeds 40.5° C (Fig.9). In addition to the copious saliva which wets the head, neck, and front legs, urine is discharged over the back legs. The function of the large bladders of desert tortoises has long puzzled zoologists, but the answer is now known. Urine is stored both as a defence against enemies, and for use in emergency cooling when the animals are unable to find shady hiding places.

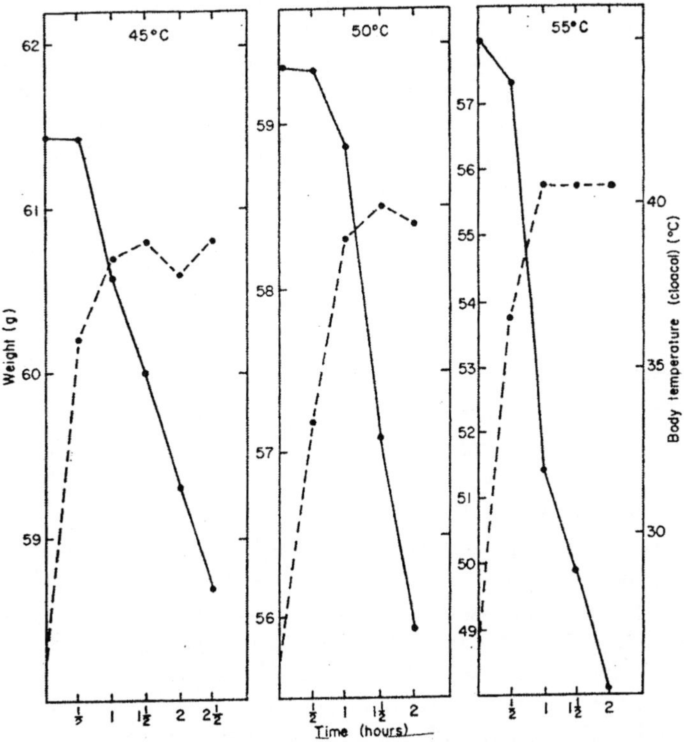

Fig 9. Loss in weight and increase in body temperature during consecutive exposures to 30 min in dry air at various high temperatures in Geochelone sulcata, weight (———); body temperature. (----). (From Cloudsley-Thompson, 1977).

Desert rodents do not sweat* - indeed it seems likely that the general absence of sweat glands in small mammals results from the necessity to conserve water that is imposed by their relatively large surface area. In order to maintain a constant, normal body temperature when the ambient temperature of the air reaches 40° C, a kangaroo-rat or jerboa would have to lose 20 per cent of its body weight per hour. In emergencies, however, a copious flow of saliva is produced in response to heat stress. This soaks the fur of the chin and throat, providing temporary relief when the body temperature

* The Nile rat (*Avicanthus niloticus*) is said to have a few sweat glands.

approaches lethal limits. Similar responses occur in ground squirrels and other small mammals whose water resources are insufficient to provide active temperature regulation for any length of time. An emergency reaction of this type may mean the difference between life and death to an animal temporarily unable to retreat into its burrow.

Larger mammals – antelope, donkeys and camels, for example – are able to sweat although, as discussed in the following section, they are also able to tolerate hyperthermia, thereby reducing water loss to a minimum.

Tolerance of heat and cold by plants and animals

Tolerance of heat by plants and animals is not unlimited, but is greater among desert organisms than in their relatives from more temperate climes. Desert plants compensate to some extend for their lack of mobility – which prevents them for moving from an unfavourable microclimate – by having an extended range over which their metabolic activities function efficiently. Their resistance to heat is partly due to increased molecular stability, particularly of enzymatic proteins whose bonding is strengthened by reduction in the degree of hydration. This phenomenon makes it possible for lichens to grow on substrates whose temperature may exceed 70° C (Hadley, 1972).

In a comparable manner, the upper lethal temperatures of desert arthropods and reptiles are sometimes higher than those of species from other terrestrial biomes. Among ants, for instance, a rough correlation exists between environment and thermal tolerance, with desert species at the upper end, and many species from cold temperate forests at the lower. The most extreme thermophiles belong to the genus *Cataglyphis* (Formicinae) in North Africa and Eurasia and *Ocymyrmex* (Myrmicinae) in southern Africa. The two are ecological equivalents (Chapter VI) of one another and are specialized in similar ways for foraging on the hot desert surface. *O. barbiger* forages in the sun on soil surfaces up to 67° C, while *C. fortis*, likewise, has the desert terrain virtually to itself at midday.

The upper lethal temperatures of desert arthropods and reptiles, likewise, are somewhat higher than those of species from other terrestrial biomes. For instance, the body temperatures of diurnal Namib dune Tenebrionidae, which are active on sandy substrates throughout the year, range between 30 and 39°C – considerably higher than those of tenebrionids of the moist tropics. Foraging limits between 7.6 and 47° C have been established for the Sonoran desert termite *Heterotermes aureus*, and between 9 and 49° C for *Gnathamitermes perplexus*. The Namib dune ant *Camponotus detritus* has a mean preferred temperature of 35°C at 100 % relative humidity while the critical thermal maximum (C T M) is as high as 53° C when the temperature is rising 1 deg C per minute; while the sub-Saharan *Cataglyphis bombycina* can withstand 45° C indefinitely and 50°C for more than 10 minutes. Again *O. barbiger*, mentioned above, has a CTM of 51.4°C while the surface of the sand is 67°C. Marsh (1987) found that the larva of the Namib desert ant-lion *Cueta trivirgata* also tolerates high body temperatures (LT_{50} = 53.4°C). By exploiting the microclimate of its pit, however, and by digging below the surface when the heat is extreme, it can capture prey at surface temperatures between 13 and 63°C. Lethal temperatures (LT_{50}) as high as 50°C for exposures of 24 hours have been recorded in the Sudanese solifugid *Galeodes granti*, 47°C in the scorpion *Leiurus quinquestriatus*, and 43° C in the tenebrionid beetles *Trachyderma hispida* and *Pimelia grandis* (Cloudsley-Thompson, 1991).

Desert lizards tolerate much higher body temperatures than do desert snakes. Most species depend upon thermal refuges for survival during the hottest periods of the day, but desert species of *Phrynosoma* show thresholds for high-temperature behaviour patterns of 2-3°C above the thresh old for non-desert species. The maintenance of an elevated body temperature carries with it a substantial benefit in that it reduces dependence on the use of thermal refuges during the day.

The fact that birds have higher body temperatures than mammals likewise gives them a big advantage in the hot desert – it allows them to rely to a greater extent upon the dissipation of heat by conduction and radiation. The body temperatures of most bird species vary between 40 and 42°C, while the lethal temperature of desert species may actually be very high indeed – above 46.5°C for the American mourning dove (*Zenaida macroura*). The ability of this species to endure hyperthermia and extensive dehydration, combined with its ability to fly long distances, allows it to meet the demands of a desert existence although it is not primarily a desert bird.

In the preceding chapter, we saw how the camel tolerates an increase in body temperature during the day, storing heat which is dissipated at night. Addax, oryx, eland, gazelles and other large desert mammals do the same. So does the ostrich, especially when it is somewhat desiccated. In summer, the body temperature of a camel may be 34°C in the morning and reach a maximum of 40.7°C in the afternoon – the temperature at which seating begins. To raise some 450 kg through 7°C takes up a great deal of heat which is lost dring the followed night. Moreover, hyperthermia, as we have seen, enhances heat loss by conduction and radiation. At the same time, in higher ambient temperatures, it reduces the difference between body temperature and ambient temperature, thereby lessening the inflow of heat from the environment.

Yet another adaptation of camels and, to a lesser extent, of donkeys is to be able to drink enormous quantities of water very quickly without suffering from water intoxication. They can tolerate rapid dilution of the blood and tissue fluids which would kill most other mammals. Finally, the rate of urine flow is low in camels and little water is lost in the faeces. Thus, a number of modifications to the general mamalian physiology are incorporated in the adaptations of camels to life in arid regions. These involve a diversity of characters which combine to make the camel uniquely successful (Schmidt-Nielsen, 1964).

Reduced metabolism and food shortage

Because water and plants are scarce, shortage of food, except at the time of the rains, is a perennial aspect of the lives of most desert animals. Arthropods frequently show surprising resistance to prolonged starvation. Scorpions, for example, are notorious for their powers of abstinence while tenebrionid beetles can often survive years of starvation. A specimen of *Blaps requieni* from Tunisia survived for five years in the laboratory without feeding

(Cloudsley-Thompson & Chadwick, 1964). Not only can scorpions withstand higher temperatures than any other arthropods, apart from Solifugae, but their rates of water loss are lower than those of other arthropods with the exception of certain Namib desert pseudoscorpions. Their metabolic rates are amongst the lowest in the animal kingdom – an order of magnitude lower than those of insects. Such low energy needs, in concert with the highest production efficiencies recorded in the animal kingdom, enable scorpions to function normally under chronic and intermittent food stress (Polis, 1991).

The capacities of all reptiles for aerobic metabolism are distinctly limited compared with those of birds and mammals of comparable size. There is little diversity in resting metabolic rates and this metabolic homogeneity, which contrasts with the diversity of basic metabolic levels found among birds and mammals, evidently pre-adapts all reptiles to life in arid regions where food is scarce. Like other ectotherms, a much greater proportion of assimilated energies is used in production than in endotherms. Even in pocket mice (*Perognathus* spp.) however, seasonal torpor results in considerable saving of energy in comparison with other species of similar body size, and it seems probable that a low metabolic rate and reduced food intake are characteristic of a majority of desert animals.

The discussion in the present chapter suggests that, with two notable exceptions, desert organisms are rather less able to tolerate adverse conditions than they are to evade or avoid them. The exceptions are poikilohygric cryptograms and chironomid midges that survive after their tissues have been completely dehydrated.

5

DETERRENCE OF HERBIVORES AND ESCAPE FROM PREDATORS

Desert plants probably interact with one another in several different ways, but these are not well understood. Although the existence of allelopathic toxins (amensalism) has not been definitely established for instance, they may well occur. Again, competition almost certainly takes place between roots for moisture, but few excavations, tracer studies or field competition experiments have yet been conducted. The requirements for shade of some annuals and perennials appear to be nonspecific and non-biotic, while very little information is available on the biology or ecology of epiphytes and parasites. Interactions between plants and animals are, however, very much better understood (Goodall *et al.*, 1981).

Many authors have commented on the high proportion of predatory species in desert faunas. This impression comes in part from the high density of arachnids and reptiles, and from the fact that a large proportion of desert birds are carnivorous. It is uncertain, however, whether this is a characteristic of all arid regions or whether it is the indirect result of deserts occurring mainly at low latitudes. A considerable amount of evidence suggests an increase in the proportion of carnivorous species in the fauna nearer to the equator than to the poles.

Seed harvesters

Food, like water, is usually in short supply in the desert, but accumulations of the seeds of ephemeral plants in favourable localities do not invariably result in outbursts of vegetation at the time of rain. This is because these accumulations are often harvested more thoroughly by ants and other seed-eating animals than are seeds scattered at random over the desert surface. In the Negev highlands, more than 50-70% of annual seed productions is collected during the summer by an ant population with a density of 800,000-1,000,000 ants per ha. (Evenari *et al.*, 1982).

The damage caused by seed-collectors is reduced by the production of numerous small seeds, as in *Schismus arabicus*, or by the evolution of larger, well-protected seeds. Another method is by minimizing the time between seed dispersal and germination, as in *Blepharis ciliaris*. Perhaps the best strategy is for seeds to germinate from the protective cover of the dead mother plant. This occurs in *Gymnorrhena micrantha* and *Erex spinosa* whose large, subterranean seeds have a better chance of completing their life cycles after germination than have the smaller aerial seeds (Gutterman, 1983).

The rate of consumption by different taxa of granivorous animals is influenced by the sizes of seeds. Seed-eating rodents tend to consume more relatively large seeds (> 1 mg), while seed-foraging ants collect more smaller seeds (< 1 mg). When rodents are removed, the densities of annual plants with large seeds, such as *Erodium texanum, E. cincutarium* and *Lotus humistratus,* may increase by as much as three orders of magnitude (Brown *et al.*, 1986). Moreover, with smaller seeds, ants can significantly increase the evenness in numbers of species. Grazing by cattle, sheep and goats may also reduce the diversity of annual plants, even when it does not affect their abundance. Plant density affects competition, growth rates, and the reproduction of desert annuals, while the inhibition of seed germination by established plants also has a competitive effect. The short-term impact of competition is much less apparent than it would be if dormancy were not an option. Further information is given in Polis (1991).

Harvesting of seeds may also affect the harvester. An additional major caste has been evolved among harvester ants. These majors use their massive, blunt-edged mandibles to strip the coat and break apart the endosperm of the seeds upon which the ants feed. They are more efficient at this than are the smaller and weaker minor workers. Milling of seeds occurs in *Pogonomyrmex badius, Solenopsis germinata, Acanthomyrmex* spp. and some granivorous species of *Pheidole.* When, however, seeds form the sole, or principal item of diet, polymorphism is much less marked. This is because all the workers are then adapted for milling (yet remain generalists as far as other tasks are concerned). Examples are afforded by *Monomorium* and *Veromessor* spp.

Avoidance of predatory enemies

Some desert animals are able to survive in desert regions because they can live where there is no water and, consequently, few predators. The addax antelope (*Addax nasomaculatus*) is a well-known example. Compared with the scimitar-horned oryx (*Oryx dammah*), which never penetrates far into the desert, it is relatively defenceless. In contrast, when attacked, the oryx charges with head down, thrusting its sword-like horns to either side with a scything movement and emitting a loud vocal challenge through its nostrils.

There must be many other examples from various animal taxa. Some 22 species of Saharan ants, of which *Messor aegyptiacus* is the most important, are able to live far from any source of moisture; but only *Cataglyphis bombycina* can inhabit moving dunes, and this is the only species not widely distributed away from the deserts. Unlike some solitary species, no social insects exhibit diapause. Indeed, the only characters distinguishing Saharan ants from those found elsewhere are the greater impermeability of their cuticles, their high thermal preferences, and high lethal temperatures (Delye, 1968). Ants prosper in the desert by exploiting favourable microhabitats where a minimum of food is obtainable throughout the year, and by the relative absence of predators in the regions.

Behavioural diversification is found among desert–inhabiting ants and their predators and parasites. Phorid flies of the genus *Apocephalus* attack the major soldiers of *Pheidole dentata* when they emerge to defend food baits from *Solenopsis texana.* Neither the minors of *P. dentata* nor the soldiers of *S. texana* are parasitised by the flies, but their influence in stampeding *P. dentata* majors tips the balance in favour of *S. texana* which then dominates the food source. In early spring and autumn, when Phoridae are absent, *P. dentata* largely replaces *S. texana* at food baits.

Adaptive Coloration in Animals

Coloration is a phenomenon so important in the lives of animals that, in almost every case, it has some kind of adaptive significance such as crypsis, communication, aposematism or optimizing radiation relationships. The diversity of colours, so apparent amongst the fauna of less open environments, is markedly absent in the desert. Here, the majority of animals are either cryptic or black. I have argued elsewhere, and at some length (Cloudsley-Thompson, 1979), that the function of the black colour of desert animals as unrelated as tenebrionid beetles, desert bee-flies (Bom-

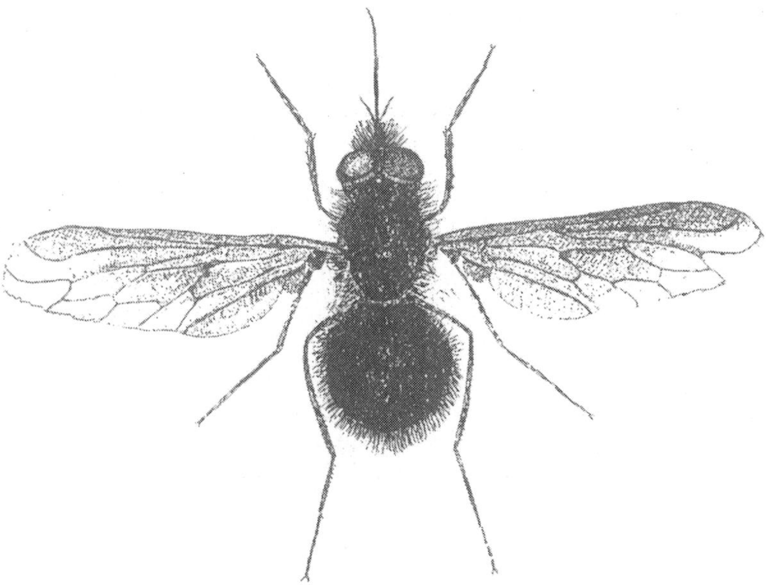

Fig. 10 : Desert beefly (Bombyliidae) (From Cloudsley-Thompson, 1991 after P.A. Buxton).

byliidae) (Fig. 10), many lizards and birds such as crows, is that it makes them conspicuous.

Colour change in reptiles is correlated both with concealment and with thermoregulation. The ability to reflect light is more marked in desert lizards than it is in species from other biomes, suggesting that colouration does have some thermal influence, but the operation of concealing and thermoregulatory colouration are largely synergistic. When the two tend to produce opposing effects, the one possessing the greater survival value is selected (Cloudsley-Thompson, 1991). Hamilton's (1973) 'maxithermy' hypothesis, that black colours enable their owners to maximize absorption of solar radiation during the early morning and late afternoon, thereby extending the time for surface activity and foraging, has engendered considerable discussion (reviewed in Cloudsley-Thompson, 1991), but is now generally discarded.

The 'maxithermy' hypothesis suggests that the high body temperatures of desert beetles are sustained because they lead to increased rates of various biological processes and are therefore advantageous. In a recent experiment to test this hypothesis, Ward (1991) found no significant difference in the environmental temperatures selected by food-deprived and fed *Onymacris plana, O. rugatipennis,* and *Physadesmia globosa*. Nor was there any significant difference among the species in the temperatures selected. A more promising alternative hypothesis, as far as the Tenebrionidae of the Namib desert are concerned, is that of Seely (1985) who suugested that these beetles – which do not appear to be distasteful as are the tenebrionids of the Great Palaearctic and American deserts – avoid many nocturnal predators by being day active, and are able to withstand higher ambient temperatures than are diurnal predators. In addition, as their metabolic rates increase, they become both faster and more adept at escaping from their enemies.

Spines, venoms and repellants

Although spiny plants are found throughout the world, spines are especially characteristic of the flora of desert regions. They deter large herbivores, mainly by causing pain, the mouth, nose and eyes, being especially sensitive. Some spines are reinforced by venoms, others damage the tissues by puncturing, ripping or

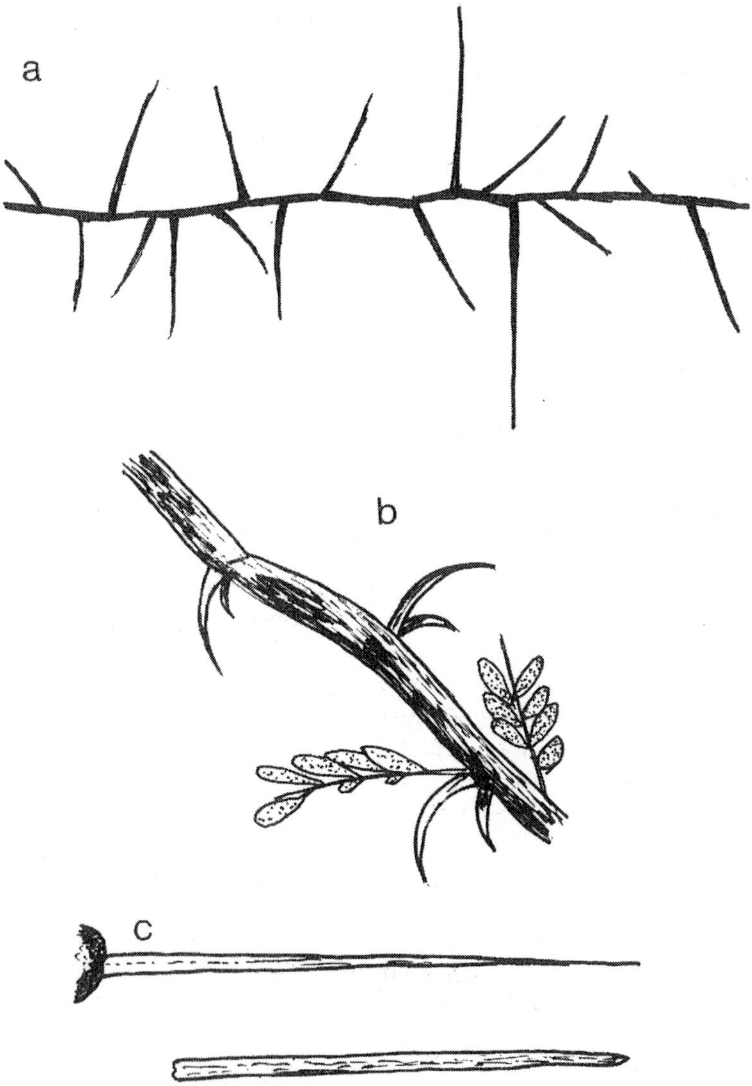

Fig. 11 : Spines of desert plants: (a) Piercing spine of *Zizyphus obtusifolia*; (b) Recurved spines of ironwood (*Olneya tesota*); (c) Cactus spine from *Opuntia phaeacantha* with the sheath removed.

piercing them (Fig.11a). Recurved spines such as those of the wait-a-bit thorn (*Acacia mellifera*) of the Great Palearctic desert, the Namib camel thorn (*A. erioloba*), the North American catclaw acacia (*A. greggii*) and ironwood trees (*Olneya tesota*) impale the flesh of moving animals. Their curvature causes the spines to penetrate deeper as the animal pulls away (Fig. 11b).

The 'spinifex' grasses (*Triodia* and *Plectrachne* spp.), mentioned in Chapter I, are perennial evergreen tussock grasses whose leaves are very harsh and spiny. The word 'spinifex' is also applied to the grass *Zygochloa (= Spinifex) paradoxa,* widely distributed in the desert areas of Australia. True spines are stiff pointed processes, branched or non-branched. Depending upon the organ from which they originate, a distinction is made between shoot spines, stipule spines, leaf spines, root spines, and so on. Spines are prominant in desert regions where vegetation is sparse and browsing pressure therefore especially heavy. Many desert Cactaceae and Euphorbiaceae are exceptionally spiny. Plant spines are a useful defence against large vertebrate herbivores, causing tissue damage in several ways: punctures, rips or tears. They may possess barbs which cause the spine to be retained in the flesh. They may also be associated with venoms and can engender infection.

Some desert plants are protected from herbivorous mammals, not by thorns and spines, but by distasteful chemicals in their tissues. The leaves of creosote bushes (*Larrea tridentata*) have a pungent smell, the pods of *Cassia senna* contain a strong purgative, and plants such as *Euphorbia* spp. and Asclepiadiaceae contain poisonous or irritant latex. The Sodom apple (*Calotropis procera*) is eaten by the grasshopper *Proeciloceros hieroglyphicus* which sequesters its poisons, but is avoided by other insects, goats and camels, while terpenoids and other aromatic compounds are produced seasonally by certain desert shrubs at the time of the rains. These are responsible for the characteristic scents of frankincense and myrrh and deter herbivorous insecticides. These include *Balanites aegyptiaca,* the mesocarp of whose fruit is lethal to mosquito larvae and, therefore, presumably to other insects, and the neem or margosa tree (*Melia azadirachta*).

Animal venoms are defensive against vertebrate predators again by causing them pain, damage to their tissues, infection, and sometimes allergic reactions. Unlike plants, animals are able to move about and can frequently direct their defences towards the enemy. It is not surprising, therefore, that animals have tended to evolve venoms rather than spines for defence. There are indeed some animals such as hedgehogs which are spiny, while lasiocampid caterpillars, such as *Gonometra postica* of the Namib, are equipped with urticating hairs attached to their integuments; but this kind of venom apparatus is less common then mobile types.

The most poisonous desert animals include vipers and cobras and other snakes of many species, gila monsters (*Heloderma suspectum*) and beaded lizards (*H. horridum*). Since these feed mainly upon eggs, their venom has primarily a defensive rather than an offensive function. Venomous invertebrates of the desert include scorpions and spiders, centipedes, Lepidoptera, ants, bees and wasps. The use of poison for defence is a simple concept, but the ways in which poisons are secreted or sequestered and delivered are extremely numerous and diverse. Unlike the secretions of repugnatorial glands venoms that are injected by teeth or stings are usually ineffectual when merely applied to the skin or taken into the alimentary canal. They are not specific to the desert fauna and, although many desert animals are markedly poisonous, the difference between them and their relatives of more humid regions is merely one of degree.

Predatory animals are often deterred more by repellant chemicals – flavours and smells – than by lethal venoms. Indeed, a venomous snake or spider may be quite edible once it has been killed, whereas a dead stink-beetle is quite as repulsive to a predator as a living one. So it is not worth attacking it in the first instance as there would be nothing to gain. Numerous desert insects are equipped with defensive glands. *Eleodes* spp. (Tenebrionidae) of the Great American desert stand on their heads when annoyed (Fig. 12) and spray the enemy with repugnatorial secretion containing benzoquinones. This has been found to repel a diversity of predators most effectively. No defence is perfect, however, and grasshopper mice (*Onychomis* spp.) learn to subdue these darkling beetles by holding them head upwards and forcing

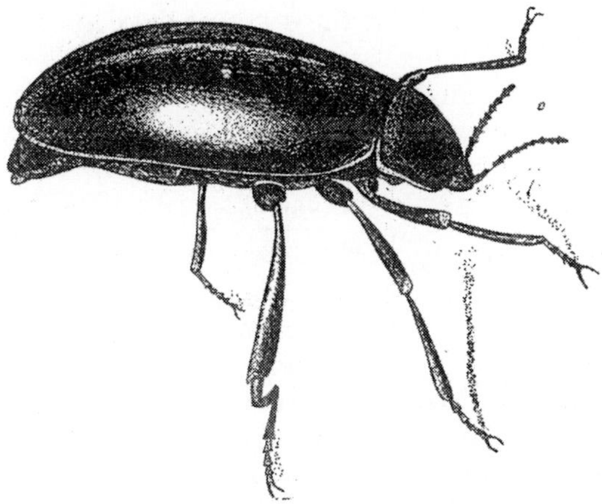

Fig. 12 : *Eleodes* sp. (Tenebrionidae) spraying its defensive secretion. (From Cloudsley-Thompson, 1991 after W. Wickler).

the tip of the abdomen into the soil where the offensive secretion is harmlessly discharged.

Several species of blister beetles (Meloidae) occur in arid regions where their triungulin larvae feed upon the eggs of locusts. *Mylabris ligata* is a common species in the northern Sudan. Other arthropods that rely for defence upon glandular secretions include whip-scorpions (e.g. *Mastigoproctus giganteus*) which protect themselves by discharging from the anal region a cloud of vapour droplets having a strong odour of acetic acid. In other species of Thelyphonida the smell resembles that of chlorine.

In addition to their venomous bites, some desert centipedes such the Californian species *Scolopendra heros*, make tiny incisions with their numerous feet. Into these they are able to drop some kind of poison that causes intense irritation so that the affected part becomes inflamed and the two rows of punctures show white against the flesh. Like nearly all Diplopoda, desert millipedes not only have a tough integument which protects them from their enemies, but they possess repugnatorial glands that secrete irritant exudates which can be discharged in the form of a fine spray. This has a strong caustic action and contains hydrocyanic acid, iodine and quinine.

Desert tortoises urinate on the sand before, during and after oviposition. This camouflages the scent of the eggs and is sufficiently unpleasant to repel some egg predators. When roughly handled, the horned lizard *Phrynosoma solare* squirts blood from its eyes while *Diplodactylus elderi* (Gekkonidae) exudes a viscous exudation from its anal region and *D. williamsi* can eject a repugnant fluid for distances of up to 0.5 m (Cloudsley-Thompson, 1991).

Interactions frequently evolve between poisonous animals and their predators. Harvester ants in the genus *Pogonomyrmex* have the most toxic venom, as far as mammals are concerned, of any known arthropods. These large ants dominate many of the desert regions of North America : on account of their formidable defences, they have relatively few predators. Their strong mandibles are used effectively against many potential arthropod predators whose limbs and appendages are easily severed. They can also be locked onto larger predators such as lizards or grasshopper mice, against which the sting is extremely effective. When an assailant is attacked, the mandibles anchor the harvester ant, affording the leverage for it to drive its sting into the adversary. In addition, the exoskeleton is hard and slippery. When individual members of a colony are threatened, they release alarm pheromones. These contain mandibular gland components as well as some of the compounds produced by the venom glands, and recruit other members of the colony to mount a mass attack.

Various species of reptiles, as well as wasps, spiders and birds, have been reported as preying on *Pogonomyrmex* spp.; but the most important predators, without doubt, are horned lizards of the genus *Phrynosoma*. These slow, broad-bodied lizards have evolved the large stomachs and digestive tracts that are necessary for digesting bulky chitinous prey and are also exceedingly resistant to the venom of harvester ants. It is evident that harvester ants and horned lizards have evolved, in parallel, as prey and predators respectively (Schmidt 1989).

Commensal interactions and coevolution

Coevolution may be defined as the simultaneous evolution of two or more taxa with close ecological interactions and, more frequently, as independent evolutionary interactions between plants and animals. These interactions are usually mutually advantageous, or at least minimise the disadvantages to which one or both associates may be exposed. Comparatively few examples of coevolution have been discovered in deserts, mainly because of the relative importance of abiotic interactions (Louw & Seely, 1982). An interesting examples is provided by the moth *Tegeticula (= Pronuba) yuccasella* and the yucca plants of the deserts of the south-western United States.

This small, variable species does not feed – the nectar secreted by the yucca flowers serves to attract other insect visitors away from the ovaries while fertilization is effected only by the moth. This enters the flower, climbs up one of the stamens and bends its head closely over the top of the antler. Uncoiling its proboscis over the anther, the yucca moth then scrapes the pollen into a lump with its maxillary palps and grasps this with its forelegs. After collecting pollen from three or four stamens, the moth then flies to another flower and investigates the condition of the ovary. It is apparently able to tell – presumably by means of pheromones – whether the flower is mature and whether eggs already been laid in it. It the new flower is suitable, the moth bores into the ovary with its ovipositor and lays its eggs, one in each of the three ovaries. The stigmas are united to form a tube down which pollen is forced by the industrious moth. Since unpollinated flowers soon die, the behaviour of *T. yuccasella* ensures a supply of food for the larvae.

Angiosperms and ants have been associated throughout most of their respective histories, and intricate coevolution of the two groups has taken place. Many plants have come to depend upon ants and other insects for pollination, and many species of ants subsist upon pollen, nectar, and seeds. In addition, however, desert bull's horn acacia trees are frequently protected by ants and, in return, have evolved holow thorns (Fig. 13) which shelter their colonies. New World species of *Acacia* have evolved beltian bodies. These are food stores produced on the tips of pinnules and rachises. Foliar nectaries also promote the welfare of the ants that inhabit acacia thorns. It has been shown experimentally that myrmecophytism protects the trees from insects such as coreid and membracid bugs, beetle larvae, leaf-eating caterpillars and so forth. As the ants patrol the leaves and branches of an acacia, they attack

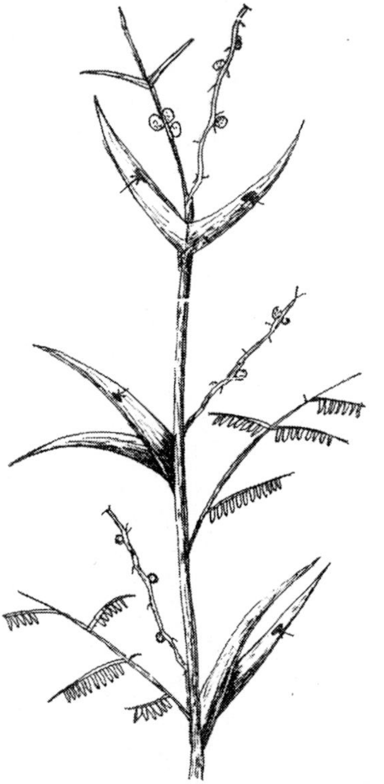

Fig.13. Bull's *horn* acacia with hollow thorns inhabited by ants.

plant-feeding insects and probably also browsing mammals. In the case of *Pseudomyrmex* spp., the workers are extremely aggressive toward intruders of all sizes and, if the tree is brushed against or shaken, rush out and attack immediately. Their stings are very painful to humans, causing a lasting burning and throbbing of the area affected.

In the coevolutionary race between predators and prey, nowhere more evident than in tropical regions including deserts, every defensive device has to be paid for in energy that might otherwise be channelled into reproduction. Predators may be either specialists or generalists. Specialists can often be distinguished by reproductive patterns that enable their population growth to parallel that of their particular species of prey. Thus they can adjust their numbers rapidly to take advantage of sudden increases in the population of the prey, and also recover speedily after setbacks when prey populations have declined.

In passing, it might be mentioned that animals show many different adaptations in responce to the pressure of selection in different environments and habitats. At the same time, similar tasks may be carried out by very different designs. No animal can complete its life cycle without feeding, for vital processes require a source of chemical energy. Some exploit other living organisms to assist them in this. With the exception of termites, whose intestines contain protozoans that digest cellulose, little is known about the gut symbionts of desert detritivores. Investigation has, however, been carried out of the gut-inhabiting bacteria and fungi of woodlice, millipedes, crickets and tenebrionid beetles. These reduce cellulose in the diet of detritivores such as *Hemilepistus* spp. (Isopoda), *Orthoporus ornatus* (Diplopoda), *Eleodes hispilabris* and *Lepidochora argentogrisea* (Tenebrionidae) (Taylor & Crawford, 1982).

Desert plants and animals are adapted to withstand the attentions respectively of herbivorous and predatory enemies. Seed harvesting is countered by adjusting the balance between large and small seeds. Predatory enemies are avoided or deterred by spines, venoms and repellents, as are herbivorous animals by plants. Further protection may be achieved by commensal relationships such as that between bull's horn spines and the ants whose welfare they promote.

6

THE DIVERSITY AND UNIQUENESS OF DESERT ORGANISMS

Despite, or rather, on account of the extreme variability of the desert climate, the desert is a remarkably homogeneous environment on a large scale, and the problems it poses to plants and animals are basically similar *viz.* how to survive at high temperature when water is in short supply and, secondly, how to compete successfully in an open, exposed environment. These are the topics to which the previous chapters of this book have been devoted. In the present chapter we should consider further, two problems referred to in the Introduction (a) to what extent can desert plants and animals be regarded as being especially diverse? (b) have they evolved unique adaptations?

Evolutionary parallelism

Living organisms that are quite unrelated often respond to environmental challenges in remarkably similar ways. The leaves and stems of desert plants, for instance are frequently oriented so that the heat load from solar radiation is reduced. The parallel with the orientation of termite nests has already been drawn (Chapter III). Locusts orient their bodies at right angles to the sun's rays in the morning and evening. At midday, however, when the heat is great, they turn their heads towards the sun, thereby reducing the surface exposed to radiation. Ostriches, camels, springbok, and probably many other species of antelope, do the same. So do reptiles, and plant leaves when they wilt as a result of excessive transpiration. The dense spination of some cacti functions like a camel's hairs, the feathers of an ostrich, or the scales of certain lizards. Indeed, feathers may well have evolved from enlarged reptilian scales and their primary function could well have been to shade the skin.

Many other examples of parallelism could be cited. Before taking flight, hawk moths (Sphingidae) raise the temperature of their thorax by 20° C above the ambient by rapidly flapping their wings and generating heat metabolically. This ensures optimum flight efficiency. The spadix of the voodoo lily (*Sauromatum guttatum*) may also reach 15°C above the ambient metabolically. This volatilizes the amines responsible for the horrible smell which attracts the flies that pollinate and flowers. When they are incubating their eggs, female Indian pythons warm the clutch by physiological thermoregulation, rapidly flexing their body muscles when ambient temperatures fall below 33°C (Cloudsley-Thompson, 1971). Again, termite nests are warmed by metabolic heat produced in their fungal gardens. One could go on almost indefinitely.

When quite unrelated animals come to look alike as a result of parallel evolution in different geographical regions, they may be known as 'ecological equivalents'. Some of the best known mammalian examples include the delightful fennec fox (*Fennecus zerda*) of the Sahara which has a number of adaptive characteristics that parallel those of the American kit foxes (*Vulpes velox* and *V. macrotis*) (which are more common in America's semi-arid plains than in her dry deserts). They are both much smaller than their relatives of temperate climes; they have well-developed sense organs, large eyes and ears, and spend the day in deep burrows thus avoiding extremes of heat. Similarly, there is a close structural similarity between the kangaroo rats (*Dipodomys* spp.) of North America and the jerboas (*Jaculus* and *Dipus* spp.) of the Old World deserts, the jack rabbits of North America, the hares of the Great Palaearctic desert and the marsupial quokka (*Setonix brachyurus*) of Australia. The sousliks (*Citellus* spp.) and prairie dogs (*Cynomys* spp.) of the New World deserts are similar to the ground squirrels (*Xerus* spp.) of the Sahara, and so on. Quokkas differ from most smaller placental mammals, however, in that they sweat freely.

Turning now to reptiles, we find marked parallels between the iguanid lizards of the American deserts and Palaearctic lacertids. Again, many members of the Agamidae occupy similar or equivalent ecological niches to those of the Iguanidae.

Examples of ecologically equivalent species include *Uromastix* spp. (Agamidae) of the Great Palaearctic desert and North American *Sauromalus* and *Cachryx* spp. (Iguanidae). The southeast Asian agamid *Liolepis* spp. resembles the North American iguanid *Amphibolurus pictus*, as does *Dipsosaurus dorsalis* in the Californian desert, while the Iranian agamid *Phrynocephalus mystaceus* and the African gecko *Geckonia chazaliae* also have many characters in common (Cloudsley-Thompson, 1991).

A dramatic example of ecological equivalents is provided by *Moloch horridus* (Fig.14a) the thorny devil of central and western Australia, a cryptically coloured and heavily armoured agamid lizard that feeds on ants, and its ecological analogue *Phrynosoma platyrhinos* (Fig.14b) the horned iguanid lizard of the North

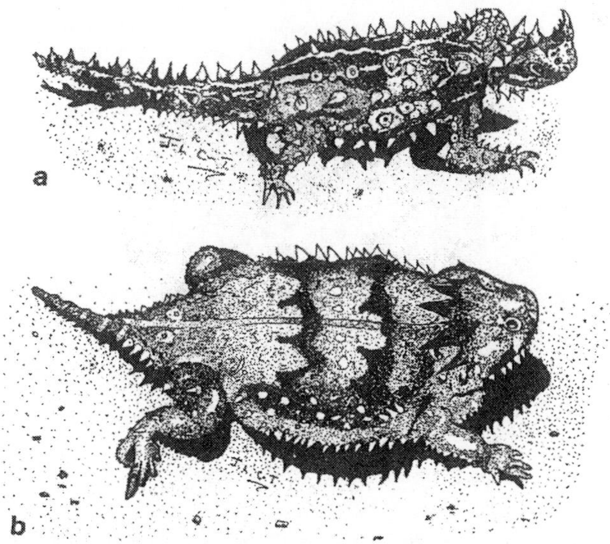

Fig. 14 : *Moloch horridus* (Agamidae), the thorny devil of the Australian deserts (b) *Phrynosoma* sp. (Iguanidae), the horned toad of the North American deserts. (From Cloudsley- Thompson, 1991).

American deserts which likewise exploits a diet of ants. Morphological analysis demonstrates that *M. horridus* and *P. platyrhinos* are anatomically closer to one another than either species is to any other member of its own lizard fauna.

Another well-known reptilian example of ecological equivalents is provided by the North American sidewinder rattlesnake (*Crotalus cerastes*, Fig 15a) and its counterpart, the Saharan sand

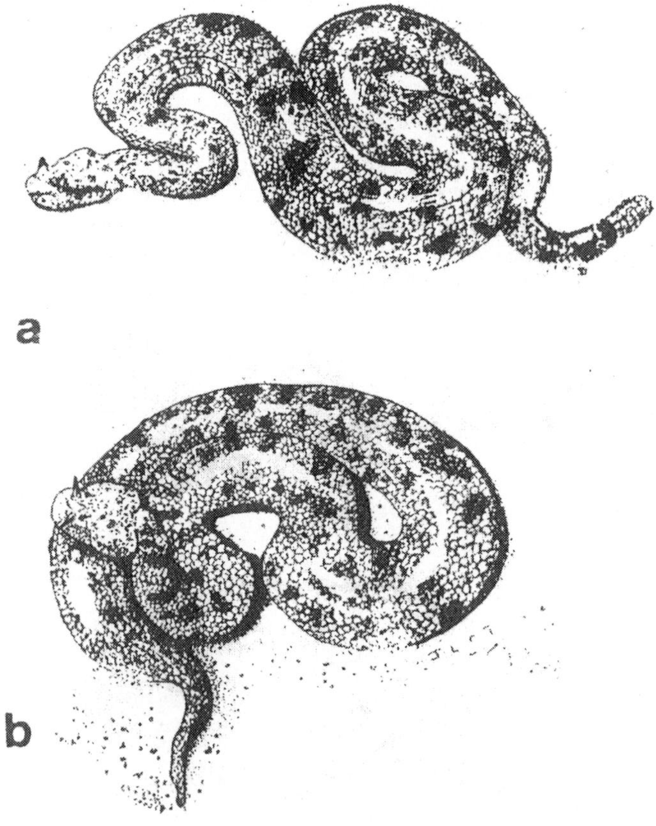

Fig. 15: (a) *Crotalus cerastes* (Crotalinae), the American sidewinding rattlesnake. (b) *Cerastes cerastes* the side- winding horned viper of the Great Palaearctic desert. (From Cloudsley-Thompson, 1991).

viper *(Cerastes cerastes*, Fig.15b). Both species move by sidewinding and hide by burying themselves in the sand. They are so much alike in appearance that, but for the rattle of *C. cerastes* and the pit between the eye and the nostril, the two species might easily be confused. Arthropodan examples of convergence are less marked than vertebrate examples, possibly because the former are smaller, more diverse, and therefore occupy less generalized ecological niches (Cloudsley-Thompson, 1991). It is easy to over-emphasise the significance of convergence, and the biologist must be careful not to allow it to bias his judgement. Nevertheless, the concept is a useful one in a restricted sense. At the same time, the hypothesis that such adaptations are correlated with the environment is supported by these extraordinarily close resemblances between unrelated species.

In Chapter IV, mention was made of the fact that ants of the genera *Ocymyrmex* and *Cataglyphis* are ecological equivalents in the deserts of southern Africa and North Africa and Eurasia respectively. Not only do they forage during the day for the corpses of insects and other arthropods that have succumbed to the heat, but their behaviour and physical appearance are remarkably similar. On a broader scale, honeypot ants (e.g. *Myrmecocystus, Melophorus, Camponotus, Plagiolepis* and *Leptomyrmex* spp.) which store liquid food in their crops, are not only ecological equivalents in different deserts of the world but, furthermore, parallel water storing by succulent plants. *Myrmecocystus* spp. are found only in the western United States and Mexico, *Camponotus inflatus, Leptomyrmex* and *Melophorus* spp. in the Australian deserts, and *Plagiolepis trimeni* in Natal. *Leptomyrmex* is a dolichoderine genus, the remainder are formicines. Australian aborigines dig up and eat the repletes of *Camponotus* and *Melophorus* spp.

No taxon of arthropods appears to have been investigated in the same way that desert lizards have been studied. Most desert regions are characterized by the presence of large, black tenebrionid beetles. *Eleodes armata* in North America might, for instance, be regarded as an ecological equivalent of *Blaps inflata* in the northern Sahara. Again, North American harvester ants of the genus *Pogonomyrmex* often construct low, saucer-shaped mounds around the nest entrance as do some central Australian

Rhytidoponera spp. and *Cataglyphis* spp. in the Sahara. These may serve to protect the nests from sheet flooding, as do the anti-flooding devices of Australian mygalomorph spiders. Convergence is less marked among desert arthropods than among desert reptiles, probably because the former are smaller, more diverse, and therefore occupy less generalized ecological niches.

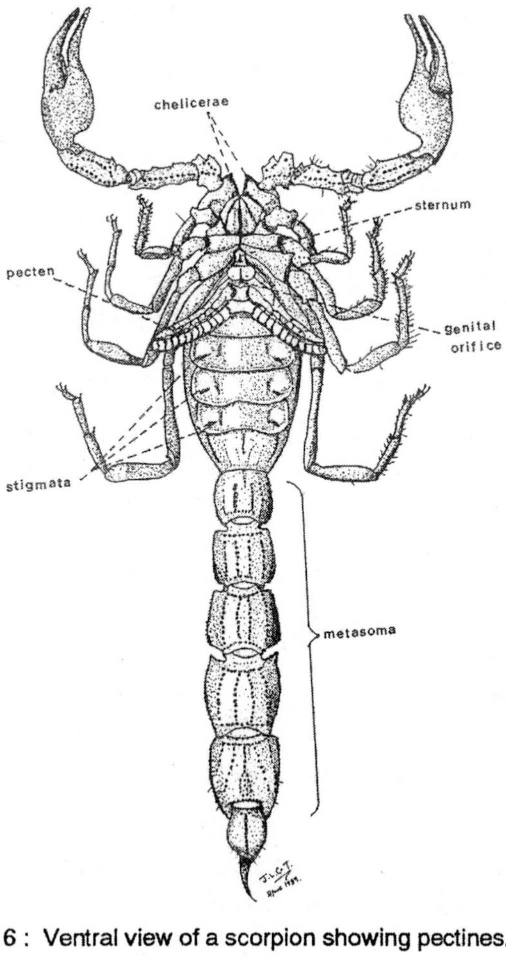

Fig. 16 : Ventral view of a scorpion showing pectines.

The diversity of desert organisms

Diversity is an imprecise concept and the answer to the question, 'are desert organisms diverse' ? is equally imprecise. In any case the question is meaningless unless qualified by reference to non- desert organisms. When compared with the flora and fauna of tropical rainforest, desert organisms cannot be regarded as being particularly diverse as regards numbers of species and their adaptations. The basic environmental problems experienced by desert plants and animals are shortage of water, daytime heat, and exposure to herbivores and predators respectively. The solutions to these problems are relatively simple and what diversity there is seems to be largely due to the fact that they have been applied by different taxa, thereby resulting in parallel evolution.

One solution to the problem has been the evolution of social behaviour. Not only do the social insects – especially termites and ants – thrive in the desert environment, but the most successful desert woodlouse (*Hemilepistus reaumuri*) survives as a result of social behaviour. Social spiders (*Stegodyphus mimoserum*) are characteristic of the Namib and Kalahari, as are social weaver-birds (*Philetairus socius*) which likewise construct enormous communal nests. These insulate the community against the excessive heat and cold of the desert environment. Subterranean rodents tend to be solitary in mesic areas but social in semi-arid regions.

Measurement of diversity

Diversity in plants and animals can be measured in a number of ways. It is often regarded as a measure of the richness of species in a community or biome. This concept provides a more useful measurement of the characteristics of a community when it is combined with an assessment of the relative abundance of the various species present. The index of diversity is a mathematical expression of the number of species, and the number of individuals of each species present. Various formulae may be used to quantify diversity. Of these, one of the most commonly is the Shannon-Wiener index. This is used when a community contains so many individuals that it can only be investigated by means of sampling. In this case, the index (D) is the ratio of the number of species to their importance as expressed, say, by their biomes or productivity within the community. ($D= \Sigma i^{s}\ \pi \log \pi$, where s= total number of species in the sample, i= total number of individuals in one species, π (a decimal fraction) is the number of one species in relation to the number of individuals in the population, and the log is to base -2 or base -e).

Methods such as this do not, however, take into account the variety of morphological forms present in the community or biome-terrestrial, fossorial, aquatic, aerial and so on. Moreover, smaller taxa not only have more species than larger ones in the same habitat, but also more individuals per species. This is simply because, on a small scale, there are many more micro- environments in the same region than there are major habitats, e.g. areas of shrubs, trees, grasses, rocks, and so on. The smaller the scale, the greater is the number of micro-habitats and, consequently, of species and individuals. Furthermore, it is a matter of opinion whether diversity should be reckoned in the number of species present or as the genetic diversity within each species.

Obviously more homogeneous environments must contain fewer species than less homogeneous environments, and the most numerous species inhabit the largest habitats. For example, *Calanus finmarchicus* has possibly one of the greatest biomasses of any animal species because it is adapted to pelagic life throughout the oceans. At the same time, because life began in the sea, there are many more marine than there are terrestrial phyla. The concept of diversity is so greatly complicated by the numerous criteria involved in its assessment that I have not attempted to define the term, any more than I would try to define precisely what is understood by the word 'species'. The reader will have to be content with a general idea as to what, in this book, is meant by both terms, because it is not possible to be precise. Genetic diversity has not been considered in these pages, nor, directly, have the numbers of species in the desert biome. I have, however, tried to give some impression of the general diversity of plants and animals with reference to the ways in which they have become adapted to withstand the stressful climatic regimes of arid environments.

Plant diversity

The vegetation of the arid zones is determined in every instance by its phylogenetic and evolutionary history. Each area has its own characteristic taxa, often related to those of adjacent mesic zones. Thus, the floristic composition of the arid regions fits into the general pattern of the distribution of vegetation throughout the world. Some desert regions lie at the borders between two or three plant geographical kingdoms. They include the Sahara and the deserts of the Middle East, the Karoo and Kalahari, and the Mexican deserts. And regions usually exhibit a gradient in floral composition which, again, corresponds with the main plant geographical world patterns.

A wide variety of growth forms occurs in arid regions, comprising the entire range from algae and lichens to annual plants and trees. Plant densities, likewise, vary widely. In consequence, the vegetation of arid lands extends between bare desert surfaces, through steppe grassland and shrubland to forested woodland. The distribution and density of the various life forms is determined by a complex of climatic, hydrographic and edaphic factors. Rainfall and freezing temperatures play a predominant role in determining gross vegetational patterns, while hydrographic and edaphic factors determine the patterns on a more detailed level. Finally, the pattern may be influenced by biotic factors such as grazing or foraging for seeds by rodents, termites and ants. Throughout the arid lands, the distribution of rainwater determines the pattern of plant growth (von Willert *et al.*, 1992).

Animal diversity

Not only do diverse organisms operate in the same way, but a single organism–even a single organ–may operate in several different ways. For example, the pectines of scorpions (Fig.16) are both mechanoreceptors and chemoreceptors. They have been shown to respond to vibrations of the ground such as those caused by the heavy footfalls of a large and potentially dangerous vertebrate, and to the delicate vibrations of the substrate caused by the approach of prey. They are also believed to respond to pheromones and carbon dioxide, as well as to the humidity of the soil and, in addition, are used to detect a suitable substrate for the deposition of spermatophores. If concentrated, carbon dioxide induces opening of the lung looks and a consequent increase of water loss by evaporation.

Many sand-dwelling desert animals are able to detect arthropodan prey using vibratory cues. Examples include antlion larvae and other predatory insects, scorpions, lizards and snakes. Both the direction and distance of a source of vibrations is detected by the North American scorpion *Paruroctonus mesaensis* by means of delicate tarsal hairs. Direction is perceived by sensing which of the legs is stimulated first, which distance is registered either by attenuation of the signal or by appreciation of the delay between stimulation of the tarsal hairs by compressional waves and the stimulation of slit sensilla by slower-moving, surface or Rayleigh waves (Brownell & Farley, 1979).

The trichobothria are sensitive to air movements engendered by the approach of prey, and the pattern of their distribution is highly adaptive. Moreover, motionless prey animals which have been stung and manage to escape before dying, are detected by chemoreceptive sensilla on the tips of the chelicerae, the tarsi of the walking legs, and also by the pectines whose sensilla release specific responses of searching, seizing and feeding.

The preceding paragraphs may give some idea of the complexity of invertebrate sensory physiology, and of the way in which a single sense organ may respond to numerous diverse stimuli while a single stimulus may influence several different types of sense organs. This condition is by no means unique to desert arthropods but is particularly well illustrated by recent work on scorpions, an order of arachnids that has radiated extensively throughout the arid regions of the globe.

Attention was drawn, in Chapter V, to some of the interactions between desert plants and animals. The diversity of interactions between biotic and physical components of the environment have already been discussed. A further arachnid example is illustrated by the eresid spider (*Seothyra* sp.) of the Namib desert. This captures crawling arthropods which become entangled in the cribellate silk of its web at all hours of the night or day. It shuttles between the cool air of its burrow and the exposed surface of the horizontal web, although this may reach a temperature of 73°C on hot days and the spider shows thermal stress above 49°C.

Size discrepancy is an aspect of sexual dimorphism among the larger Mygalomorphae of the Australian desert. In particular, the small size of the male may well be an adaptation to the hazards of the environment. By shortening the time for development of the male, mortality due to extended drought and associated desiccation in relatively shallow burrows, as well as through sudden flooding, is reduced. Although juvenile females may also perish, there will always be some older females with deeper burrows that survive to maintain the population.

All these points were discussed at the XI International Congress of Arachnology (Koponen *et al.*, 1990) and provide an indication of the diversity found among desert arachnids. Many other examples could be cited in both this and other taxa.

It will have already become evident that termites and ants are abundant in the arid ecosystems of the world and the number of species present in any particular region is governed by primary production and, in the case of termites especially, by temperature. Warmer habitats have more species as, also, have areas with poorer soils. There is considerable evidence that convergent evolution has taken place in both orders. Some taxa compensate in diversity for the lack of other insect taxa, and this is especially evident with respect to granivory, a subject discussed in Chapter V. Thus, both termites and ants are important in determining the composition of the plant community in a specific area. Termites are especially important in this context as they also affect the chemical and hydrological properties of the soil, and because they also harvest and recycle plant materials. At the same time, interactions between species, especially competition, are important in structuring communities; while resources are divided according to activity patterns, differences in microhabitats, size or type of food consumed.

One might perhaps conclude that certain taxa are especially fitted for life in arid regions, and these have diversified widely in the desert. Other taxa, not particularly well suited to conditions of heat and drought, are represented by relatively few species, while many animals are not able to survive in deserts and are never found there.

Habitat selection among desert animals

Habitat selection is a means by which competition is reduced between related species, and diversity thereby engendered. It occurs throughout all the biomes of the world, including the desert regions (Polis, 1991). Among detritivores and associated arthropods, relatively low values have been obtained in studies carried out in the Great Basin desert of North America, the Chihuahuan, Namib and Kalahari deserts. These studies reveal that arthropod assemblages on the desert surface vary qualitatively and quantitatively from season to season, but species turnover has been little studied.

Anuran communities have been studied to some extent in the Chihuahuan desert by Woodward (1987) who noted that temporary and permanent ponds differ in two important respects. Temporary pools contain smaller numbers and fewer types of predators than do permanent ponds, but they also contain higher densities of anuran tadpoles. Possibly predator-prey interactions are stronger in temporary pools, and competitive interactions stronger in permanent ponds. Rather more work has been carried out on desert reptiles and it is evident that each species utilizes its food resources in a non-random fashion.

The deserts of Australia are particularly rich in lizard species. Pianka (1969) suggested the following explanations : (a) lizards userp ecological roles played by some worms, insects, snakes and mammals in the North American deserts; (b) there is greater temporal partitioning because of the milder and more constant Australian desert climate; (c) Australian desert lizards are more narrowly specialised; (d) there is greater spatial heterogeneity in the Australian deserts because of an interdigitation of other types of desert habitat. Another reason may be in the nature of spinifex grasses. These have a distinctive growth form; tightly bunched clumps of interlacing, spiky leaves serve as protection for animals small enough to enter the clumps through the interstices between the spikes. Furthermore, the microclimate within the clumps is considerably ameliorated in comparison with elsewhere. Finally, the infertile soils of Australia result in sparse and unreliable food production, a condition unfavourable for endotherms, but more suitable for ectotherms such as reptiles.

Different species of desert lizards may avoid intraspecific competition with one another, even in a comparatively simple ecological system such as the creosote bush (*Larrea tridentata*) community of the Chihuahuan desert, by temporal and spatial resource partitioning. Thus, although there is considerable spatial overlap between *Cnemidophorus tigris* and *Holbrookia texana*, no less than five different patterns of activity can be distinguished. In contrast, *Uta stansburiana, Sceloporus magister, Phrynosoma modestum, P. cornutum* and *Gambelia wislizenii* are bimodal. Individual patterns of activity may vary, however: some species are active only in the morning, others only at midday, only in the afternoon, morning or afternoon, or at midday.

The avian faunas of desert regions reflect the ability of birds to fly rapidly from one locality to another, either according to season or unpredictably, and to exercise a choice between a mosaic of arid habitats. Birds respond actively to variations in food supply. Because the patches in which food is relatively plentiful tend to be small and scattered, gregarious and colonial species tend to be favoured by natural selection.

The structure of their selected microhabitat impinges on the foraging efficiencies of desert rodents. Because many aspects of size and locomotory gait affect the detection and efficient gathering of desert seeds, morphology and habitat use are closely related. Predators are another major factor to influence size and morphology. Speed and erratic movements may deter some potential predators, but natural selection favours a range of escape mechanisms because this prevents the evolutionary specialization of predators on a single strategy of attack. Furthermore, the large eyes, excellent hearing and bipedal locomotion of kangaroo rats adapt them to open areas, whereas small, quadripedal mice are more adept at scurrying through vegetation and thereby avoiding both detection and capture. These and other aspects of habitat selection are discussed in various chapters of the recent book edited by Polis (1991).

The uniqueness of desert organisma

Finally, we come to the second of the two questions with which the present volume has been primarily concerned, *viz* are desert organisms unique and to that extent are they diverse. This was discussed at some length in Chapter I and the conclusions reached there have subsequently been confirmed. The taxa that have diversified most in desert regions – arthropods, reptiles and small mammals – are those whose ancestors were to some extent pre-adapted to life in desert environments. There are few unique specializations, but adaptations already present have often been refined to an extreme degree.

The major constraint on life in desert regions is shortage of water. Poikilohygrous plants are entirely dependent upon atmospheric moisture, and are capable of being dehydrated without damage to their cells. They include bacteria, terrestrial algae, fungi, lichens, mosses, ferns (e.g. *Cheilanthes* spp.) and a few angiosperms (e.g. *Myrothamnus flabellifolius, Vellozia elegans* and *Borya nitida*). In contrast, homeohydrous plants are relatively independent of atomospheric moisture and cannot withstand complete dehydration. Drought resistant species survive (a) by completing their life cycles before the onset of the dry season, (b) by means of roots that reach the water table (phreatophytes) or through water storage stems and roots, (c) by minimizing water loss through stomatal closure, reduction of surface area, waxy cuticles and leaf dehiscence. In arid environments, plants with a C_4 photosynthetic pathway have a higher heat tolerance than C_3 plants, whilst CAM plants use water more efficiently than either.

Animals have the great advantage of mobility which enables them to seek shade, burrow, search for food and water, and even to migrate. Many desert species withstand the dry season in-diapause, while reproduction coincides with the start of the rainy season. Water loss through evaporation is often reduced by hairs, scales, wax coatings etc., while honey-pot ants actually store liquid food in their crops. Even this is not a unique specialization to seasonal aridity, although the most extreme examples are found among ground dwelling desert species of the United States and Mexico.

Conclusions

In the Introduction (Chapter I), I argued that the relative diversity of desert plants is greater than that of desert animals because long-lived plants, unlike animals, are unable to seek out favourable microhabitats during the day and are therefore compelled to withstand the full brunt of the heat and low atmospheric humidity. Moreover, I would also argue that the entire plant kingdom – because plants are less mobile than animals and, consequently, have less evolutionary options available to them – is in some respects relatively more diverse than the animal kingdom.

In Chapter II, it was shown that ephemeral plants, mostly annuals in desert regions, evade climatic extremes and prolonged periods of drought in the form of dormant tissues – roots, bulbs and corms or, more usually, seeds. The latter are paralleled by small invertebrates which evade unfavourable conditions in a state of diapause. Thus uniformity is imposed upon a diversity of desert organisms.

Avoidance of and escape from climatic extremes, discussed in Chapter III, is the most usual adaptation to desert life, responses to heat being primarily behavoural; to shortage of water, physiological. Tolerance of extremes, discussed in Chapter IV, is an adaptational strategy less frequently adopted. Finally, adaptations concerned with the deterrence of herbivores, avoidance of granivores, and escape from predators, are outlined in Chapter V. Habitat selection is considered in the present Chapter.

Desert organisms survive by:-

(a) Evasion of drought and heat by completing their life cycle before the onset of stressful conditions, e.g. ephemeral plants with drought – resistant seeds, arthropods in diapause, etc.

(b) Avoidance and escape from heat and drought, in the case of plants either by the possession of long roots that reach the water table (phaetophytes) or by xeromorphic characters. Among animals are found morphological characters that reduce heat gain and water loss, aestivation, retreat into burrows and shelter during the day, etc.

(c)

Tolerance of heat and desiccation by poikilohygric cryptograms, certain Chironomidae etc. crassulacean acid metabolism, conservation of water, thermoregulation, etc.

(d) Deterrence of herbivores and escape from carnivores.

(e) Habitat selection.

These are the selective factors responsible for diversity and adaptation. The latter can be classified along a continuum of adaptation – from evasion at one extreme to tolerance at the other. Only a very few adaptations to desert life can be considered as being in any way unique. Whether or not desert organisms can be regarded as being diverse will depend ultimately upon the subjective view of the reader and upon any presuppositions he or she may have as to the nature of the desert environment and what might be expected in response thereto.

REFERENCES

No attempt has been made to compile a detailed bibliography of the many topics considered in the preceding pages. In general, books and reviews that afford an entry to the literature have, for reasons of conciseness and brevity, been cited in preference to original publications.

Applin, D.G., Cloudsley-Thompson, J.L. & Constantinou, C. (1987). Molecular and physiological mechanisms in chronobiology – their manifestations in the desert ecosystem. *Journal of Arid Environments*, **13** : 187-197.

Baharav, D. (1983). Reproductive strategies in female Mountain and Dorcas gazelles (*Gazelle gazella gazella* and *Gazella dorcas*). *Journal of Zoology*, London, 200: 445-453.

Batanouny, K.H. (1981). *Ecology and Flora of Qatar*. Oxford: Alden Press. xiv + 245 pp.

Belk, D. & Cole, G.A. (1975). Adaptational biology of desert temporary-pond inhabitants pp. 207-226. *In* Hadley, N.F. (Ed.). *Environmental Physiology of Desert Organisms*. Stroudsberg, Penn.: Dowden Hutchinson & Ross. viii+283 pp.

Brownell, P. & Farley, R.D. (1979). Prey-localizing behavior of the nocturnal desert scorpion, *Paruroctonus mesaensis*: orientation to substrate vibrations. *Animal Behaviour*, 27: 185-193.

Cloudsley-Thompson, J.L. (1964). Terrestrial animals in dry heat: introduction. pp 447 – 499 in Dill, D.B., Adolph, E.F. & Wilber C. G. (Eds.) Adaptation to the Environment (Handbook of Physiology Section 4), washington, D. C. : American Physiological Society. ix + 1056 pp.

Cloudsley-Thompson, J.L. (1971). *The Temperature and Water Relations of Reptiles*. Watford, Herts. : Merrow. 159 pp.

Cloudsley-Thompson, J.L. (1977). *Man and the Biology of Arid Zones.* London: Edward Arnold. ix+182 pp.

Cloudsley-Thompson, J.L. (1979). Adaptive functions of the colours of desert animals. *Journal of Arid Environments,* **2:** 95-104.

Cloudsley-Thompson, J.L. (1991). *Ecophysiology of Desert Arthropods and Reptiles.* Berlin, Heidelberg: Springer-Verlag. x+203 pp.

Cloudsley-Thompson, J.L. & Chadwick, M.J. (1964). *Life in Deserts.* London: Foulis. xvi + 218 pp.

Cody, M.L. (1989). Growth-form diversity and community structure in desert plants. *Journal of Arid Environments,* **17 :** 199- 209.

Crawford, C.S. (1981). *Biology of Desert Invertebrates.* Berlin, Heidelberg, New York: Springer-Verlag. xvi+314 pp.

Delye, G. (1968). *Recherches sur l'Ecologie, la Physiologie et l'Ethologie des Fournis du Sahara.* Thesis de l'Universite d'Aix-Marseilles. 155 pp.

Denlinger, D.L. (1986). Dormancy in tropical insects. *Annual Review of Entomology,* **31:** 239-264.

Dobson, A.P. & Crawley, M.J. (1987). What's special about desert ecology? *Trends in Ecology and Evolution,* **2:** 145-146.

Edney, E.B. (1987). *Water Balance in Land Arthropods.* Berlin, New York: Springer-Verlag. xii+282 pp.

Evenari, M., Shanan, L. & Tadmor, N. (1982). *The Negev. The challenge of a desert* (2nd Edn.). Harvard, Mass.: Harvard University Press. ix+437 pp.

Goodall, D.W., Perry, R.A. & Howes, K.M.W. (Eds.), (1981). *Arid-land Ecosystems: Structure, functioning and management.* Vol. 2, Cambridge, London, New York: Cambridge University Press. xvii + 605 pp.

Gould, S.J. & Vrba, E. (1982). Exaptation – a missing term in the science of form. *Paleobiology,* **18 :** 4-15.

Gutterman, Y. (1983). Survival mechanisms of desert winter annual plants in the Negev highlands of Israel. *Scientific Reviews on Arid Zone Research,* **1 :** 249-283.

Hadley, N.F. (1972). Desert species and adaptation. *Ameircan Scientist,* **60:** 338-347.

Hadley, N.F., Quinlan, M.C. & Kennedy, M.L. (1991). Evaporative cooling in the desert cicada: thermal efficiency and water/metabolic costs. *Journal of Experimental Biology,* **159:** 269-283.

Hamilton, W.J. III (1973). *Life's Color Code.* New York: McGraw-Hill. x+238 pp.

Hinton, H.E. (1960). Cryptobiosis in the larva of *Polypedilum vanderplanki* Hint. (Chironomidae). *Journal of Insect Physiology,* 5 : 286-300.

Hohn, R. & Petermann, J. (1980). *Curiosities of the Plant Kingdom.* (Tansl. H. Liebscher). London : Cassell. 212 pp.

Holldobler, B. & Wilson, E.O. (1990). *The Ants.* Berlin, Heidelberg, London : Springer-Verlag. xii + 732 pp.

Hudson, J.W. & Bartholomew, G.A. (1964). Terrestrial animals in dry heat : estivators pp. 541-550 *in* Dill, D.B., Adolph, E.F. & Wilber C.G. (Eds.) Adaptation to the Environment (Handbook of Physiology Section 4), Washington, D.C.: American Physiological Society. 1056 pp.

Koponen, S., Lechtinen, P.T. & Rinne, V. (Eds.) (1990). Proceedings of the XI International Congress of Arachnology, Turku, Finland, 7-12 August 1989. *Acta Zoologica Fennica,* No. **190 :** 1-422.

Lambert, M.R.K. (1984). Amphibians and reptiles pp. 205-227 *in* Cloudsley- Thompson J.L. (Ed.) *Sahara Desert* (Key Environments). Oxford, New York : Pergamon Press. x + 348 pp.

Louw, G.N. (1990). Physiological studies on the Namib fauna: a brief critique. *In* Seely, M.K. (Ed.) *Namib Ecology : 25 years of Namib research.* Transvaal Museum Monograph No. 7: 203-207.

Louw, G & Seely, M. (1982). *Ecology of Desert Organisms.* London : Longman. 194 pp.

McGinnies, W.G., Goldman, B.J. & Paylore, P. (Eds.) (1968). *Deserts of the World. An appraisal of research into their physical and biological environments.* Tucson: University of Arizona Press. xxviii + 788 pp.

Marsh, A.C. (1987). Thermal responses and temperature tolerance of a desert ant-lion larva. *Journal of Thermal Biology,* **12**: 295-300.

Oksanen, L. & Oksanen, T. (1989). Natural grazing as a factor shaping out barren landscapes. *Journal of Arid Environments* **17 :** 219-233.

Pianka, E.T. (1969). Habitat specificity, speciation and species diversity in Australian desert lizards. *Ecology,* **50 :** 498- 502.

Polis, G.A. (1980). Seasonal patterns and age-specific variation in the surface activity of a population of desert scorpions in relation to environmental factors. *Journal of Animal Ecology,* **49:** 1-18.

Polis, G.A. (Ed.) (1991). *The Ecology of Desert Communities.* Tucson : University of Arizona Press. ix + 457 pp.

Rundel, P.W. (1978). Ecological relationships of desert fog zone lichens. *Bryologist*, **81:** 277-293.

Safriel, U.N., Ayal, Y., Kotler, B.P., Lubin, Y., Olvig-Whittaker, L. & Pinshow, B. (1989). What's special about desert ecology? Introduction. *Journal of Arid Environments*, **17** : 125-30.

Schmidt, J.O. (Ed.) (1989). *Special Biotic Relationships in the Arid Southwest*. Albuquerque, N.M.: University of New Mexico Press. xiii + 152 pp.

Schmidt-Nielsen, K. (1964). *Desert Animals. Physiological problems of heat and water*. Oxford: Clarendon Press. xv + 277 pp.

Seely, M.K. (1985). Predation and environment as selective forces in the Namib Desert. *In*: Vrba, E.S. (Ed.), *Species and Speciation. Transvaal Museum Monograph*, No. **4** : 161-165.

Slobodkin, L.B., (1989). What have I learned about deserts? *Journal of Arid Environments*, **17** : 287-292.

Smith, S.D. & Nobel, S.P. (1986). Desert pp. 13-62 *in* Baker, N.R. & Long, S.P. (Eds.). *Photosynthesis in Contrasting Environments*. Amsterdam : Elsevier (Biomedical Division). 423 pp.

Solbrig, O.T. & Orians, G.H. (1977). The adaptive characteristics of desert plants. *American Scientist*, **65** : 412-421.

Taylor, E.C. & Crawford, C.S. (1982). Microbial gut symbionts and desert detritivores. *Scientific Reviews on Arid Zones Research*, **1** : 37-52.

Tocque, K. & Tinsley, R.C. (1991). The influence of desert temperature cycles on the reproduction biology of *Pseudodip- lorchis americanus* (Monogenea). *Parasitology*, **103** : 111-120.

Von Willert, D.J., Eller, B.M., Werger, M.J.A., Brinckmann, E. & Ihlenfeldt, H.D. (1992). *Life Strategies of Succulents in Deserts, with special reference to the Namib desert*. Cambridge : Cambridge University Press. xix + 340 pp.

Wallwork, J.A. (1982). *Desert Soil Fauna*. New York : Praeger. x + 296 pp.

Ward, D. (1991). A test of the 'maxithermy' hypothesis with three species of tenebrionid beetles. *Journal of Arid Environments*, **21** : 331-336.

Whitford, W.G. (Ed.) (1986). *Pattern and Process in Desert Ecosystems*. Albuquerque : University of New Mexico Press. ix + 139 pp.

Woodward, B.D. (1987). Clutch parameters and pond use in some Chihuahuan Desert anurans. *Southwestern Naturalist*, **32** : 13-19.

INDEX